VOLUME 458   NOVEMBER 1981

THE ANNALS

of The American Academy *of* Political
and Social Science

ISSN 0002-7162

RICHARD D. LAMBERT, *Editor*
ALAN W. HESTON, *Associate Editor*

TECHNOLOGY TRANSFER:
NEW ISSUES, NEW ANALYSIS

Special Editors of this Volume

ALAN W. HESTON

Professor of South Asian Economics
University of Pennsylvania
Philadelphia, Pennsylvania

HOWARD PACK

Professor of Economics
Swarthmore College
Swarthmore, Pennsylvania

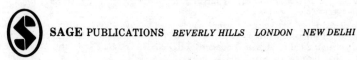

SAGE PUBLICATIONS *BEVERLY HILLS LONDON NEW DELHI*

THE ANNALS

© 1981 *by* The American Academy *of* Political *and* Social Science

PRISCILLA A. ESTES, *Assistant Editor*

Editorial Office: 3937 Chestnut Street, Philadelphia, Pennsylvania 19104.

For information about membership* (individuals only) and subscriptions (institutions), address:

SAGE PUBLICATIONS, INC.
275 South Beverly Drive
Beverly Hills, Calif. 90212 USA

From India and South Asia, write to:

SAGE INDIA
P.O. Box 3605
New Delhi 110 024
INDIA

From the UK, Europe, the Middle East and Africa, write to:

SAGE PUBLICATIONS LTD
28 Banner Street
London EC1Y 8QE
ENGLAND

*Please note that members of The Academy receive THE ANNALS with their membership.

Library of Congress Catalog Card Number 81-40889
International Standard Serial Number ISSN 0002-7162
International Standard Book Number ISBN 0-8039-1706-6 (Vol. 458, 1981, paper)
International Standard Book Number ISBN 0-8039-1017-4 (Vol. 458, 1981, cloth)

Manufactured in the United States of America. First printing, September 1981.

The articles appearing in THE ANNALS are indexed in *Book Review Index; Public Affairs Information Service Bulletin; Social Sciences Index; Monthly Periodical Index; Current Contents: Behavioral, Social, Management Sciences;* and *Combined Retrospective Index Sets.* They are also abstracted and indexed in *ABC Pol Sci, Historical Abstracts, Human Resources Abstracts, Social Sciences Citation Index, United States Political Science Documents, Social Work Research & Abstracts, Peace Research Reviews, Sage Urban Studies Abstracts, International Political Science Abstracts,* and/or *America: History and Life.*

Information about membership rates, institutional subscriptions, and back issue prices may be found on the back cover of this issue.

Advertising. Current rates and specifications may be obtained by writing to THE ANNALS Advertising and Promotion Manager at the Beverly Hills office (address above).

Claims. Claims for undelivered copies must be made no later than three months following month of publication. The publisher will supply missing copies when losses have been sustained in transit and when the reserve stock will permit.

Change of Address. Six weeks' advance notice must be given when notifying of change of address. Please send old address label along with the new address to insure proper identification. Please specify name of journal. Send change of address to: THE ANNALS, c/o Sage Publications, Inc., 275 South Beverly Drive, Beverly Hills, CA 90212.

Origin and Purpose. The Academy was organized December 14, 1889, to promote the progress of political and social science, especially through publications and meetings. The Academy does not take sides in controverted questions, but seeks to gather and present reliable information to assist the public in forming an intelligent and accurate judgment.

Meetings. The Academy holds an annual meeting in the spring extending over two days.

Publications. THE ANNALS is the bimonthly publication of The Academy. Each issue contains articles on some prominent social or political problem, written at the invitation of the editors. Also, monographs are published from time to time, numbers of which are distributed to pertinent professional organizations. These volumes constitute important reference works on the topics with which they deal, and they are extensively cited by authorities throughout the United States and abroad. The papers presented at the meetings of The Academy are included in THE ANNALS.

Membership. Each member of The Academy receives THE ANNALS and may attend the meetings of The Academy. Membership is open only to individuals. Annual dues: $24.00 for the regular paperbound edition (clothbound, $36.00). Add $6.00 per year for membership outside the U.S.A. Members may also purchase single issues of THE ANNALS for $5.00 each (clothbound, $7.00).

Subscriptions. Institutions may subscribe to THE ANNALS at the annual rate: $42.00 (clothbound, $54.00). Add $6.00 per year for subscriptions outside the U.S.A. Institutional rates for single issues: $7.00 each (clothbound, $9.00).

Single issues of THE ANNALS may be obtained by individuals who are not members of The Academy for $6.00 each (clothbound, $7.50). Single issues of THE ANNALS have proven to be excellent supplementary texts for classroom use. Direct inquiries regarding adoptions to THE ANNALS c/o Sage Publications (address below).

All correspondence concerning membership in The Academy, dues renewals, inquiries about membership status, and/or purchase of single issues of THE ANNALS should be sent to THE ANNALS c/o Sage Publications, Inc., 275 South Beverly Drive, Beverly Hills, CA 90212. Sage affiliates in London and India will assist institutional subscribers abroad with regard to orders, claims, and inquiries for both subscriptions and single issues.

CONTENTS

BOOK DEPARTMENT
INTERNATIONAL RELATIONS AND POLITICS

UNITED STATES

SOCIOLOGY

ECONOMICS

PREFACE

The articles in this volume consider the role of technology in the process of economic development. Some emphasize the interaction between advanced countries and less-developed countries (LDCs); some the interaction among LDCs, and a few the activities individual countries have taken to foster their own technological development. The "transfer of technology" is seen by almost all the authors as the implanting and nurturing of methods of production in which the critical elements are not the physical means of production, be they buildings or machinery, but the knowledge of how to operate such means effectively, or in one phrase, the achievement of technological mastery. The characteristic features of the required knowledge are that it is not easy to specify in blueprints or manuals (see the articles by Dahlman and Westphal, McCulloch, and Teece in this volume); it is difficult to negotiate about, since so much of it is tacit and thus a "fair" price is difficult to define; and that recipients of technology cannot be passive but must undertake purposive action to increase the ability to identify their needs, to learn about those technologies that might be particularly useful, and, especially, to operate them successfully. Unlike some international agencies and some LDC analysts who ritually invoke the free transfer of—often high—technology as the critical missing catalyst to accelerated development, almost every author alludes to the hard work and substantial monetary costs incurred by both transferor and recipient: no magic wands exist to accelerate development, and not only are there no free sandwiches, there are no free high-yield seeds with which to grow the grain for the bread.

An important organizing principle for much of the analysis is the distinction between the high costs—and risk—of developing a new technology and the theoretically low cost of disseminating the technology once it has been fully developed. These propositions, if correct, lead to a dilemma. Insofar as individual firms undertake research and development, some mechanism must be introduced to guarantee a reasonable rate of return on the large investment. However, such legal means as patents increase the cost of dissemination above the desirable level as those who want to use the new technology must pay royalties which may exceed the social cost of transmission.

Many of the articles—for example, those by Evenson, Dahlman and Westphal, McCulloch, Pray, and Teece—suggest that the second proposition, the low cost of dissemination, is in fact incorrect in both agriculture and industry. In industry, it is not a set of blueprints and machinery that is transferred but a substantial body of knowledge, embodied in both individuals and firms. Large costs must be incurred by either the transmitting or receiving entity. In some of the articles—for example, Teece's and Lecraw's—it is noted that given the technical difficulties and expense inherent in transferring organizational ability, and the resulting uncertainty about what constitutes a fair price for such services, many firms prefer internal transactions and thus engage in direct foreign investment. Of particular interest is the recently discerned trend that considerable transfer of technology, both in the form of adapted products and processes, is being carried out by LDC-based multinational corporations (MNCs). Though little of the knowledge so transmitted represents a breakthrough in the international frontier of knowledge, much of it is particularly useful in the recipient country—for example, refer to the article by Lecraw. Conversely, Magee offers an explanation of the perplexing fact that MNCs in the advanced countries have tended to perform little, if any, research on products or processes relevant for LDCs. He attributes this to the difficulty of appropriating the full benefits of the research since simple machines or products are more easily emulated than are the complex processes and products in which they normally specialize.

The conditions for the successful generation and transfer of appropriate technology are best seen in agriculture. High-yield seeds were developed by internationally financed and staffed research institutions, but successful local use was dependent on a complex set of domestic efforts—for example, see the articles by Evenson and by Pray—including research to adapt seeds to local climatic conditions and the determination of correct amounts of complementary productive factors such as fertilizer and water. These were expensive activities for the recipient countries, even though the basic research was internationally sponsored and no licensing fees were necessary for relevant process information, nor were high prices charged for seeds. Expenditures on these adaptive activities are shown to have generated remarkably high rates of return, as Evenson notes. A major difference between agriculture—or, more precisely, the experience with the green revolution—and industry is that the appropriability problem does not arrive in the former, as much of the R&D was financed by international agencies and this cost did not have to be recovered.

In the industrial sector there has been significant importation of plant, equipment, and the acquisition of licenses for entire processes to produce specific products. There has been limited adaptation of products, though more of processes, to suit LDC conditions. Although a large literature attests to the physical possibility and economic efficiency of adapting production methods to take account of the relative abundance of labor and scarcity of capital in LDCs, many countries and the firms within them have forgone the potential benefits of such activity. These benefits include large increases in output, employment, and profits, as my article points out, and an improved distribution of income both among families and between rural and urban sectors, as Ranis notes in his article. In addition, the inappropriate technology when chosen is often used inefficiently as local firms are unable to achieve technological competence of even a simple type (see Dahlman and Westphal in this volume). It is not an accident that countries exhibiting the lowest income inequality in international comparisons have exhibited largely correct choices about which industrial sectors to emphasize as well as the appropriate technology for each of them, as Ranis points out.

One source of the different experience in agriculture and manufacturing may lie in the different economic environments in which they typically function. Food production is of great importance in most LDCs as a shortfall would be immediately manifest to governments in high foreign exchange outlays for imported food. Avoidance of such expenditures of scarce currency has undoubtedly increased the receptivity of governments to programs designed to increase agricultural productivity. Other favorable conditions are discussed by Pray. In contrast the decision to foster manufacturing through protection from international competition has rarely led to explicit budgetary outlays, though costs are incurred by consumers in the form of higher prices as well as by workers who fail to obtain jobs because of the use of inappropriately capital intensive technology. The benefits forgone by both groups constitute public goods in which large numbers of individuals would share if they were realized and from which none can easily be excluded. Under these conditions it is difficult, as noted in my article, to organize the potential constituency for appropriate public action in the industrial sector.

The production of new technology within the LDCs themselves is the subject of many of the contributions. In recent years there have been recurrent calls for an end to technological "dependence" and a move toward autonomy by the LDCs. Whether the advocates of this approach have meant that each country should be largely self-supporting, or only the LDCs as a group, is often unclear. Paradoxically, many gathering under the banner of independence also advocate reductions in the costs of acquiring proprietary technology, particularly the reform of the patent system, but also a code forbidding a large number of existing practices, such as prohibiting exports to third countries of products based upon licensed processes. The demand for less expensive access to technology overlooks the fact that the cheaper it is to obtain licenses, the greater the incentive of LDC firms to rely on them and forgo creating a domestic capacity for technological adaptation and innovation. As Lall and others note, encouraging the domestic generation of new technology requires explicit actions by each LDC and one negative action may be the limitation of the extent to which local firms may utilize licenses. Lall catalogues the structure of incentives provided in India and derives a tentative list of its accomplishments as manifest in recent exports of Indian technology. In an overview and careful analysis of a major empirical research effort in Latin America, Teitel observes that the conditions propitious to LDC-based technical innovation are elusive: neat analytic frameworks that emphasize one causal feature, whether demand conditions or the form of ownership or the extent of competition, do not capture the complexity of recent experience.

Lall, Lecraw, and Teitel document the many innovations—most often small changes in process technology—that seem to characterize such advanced LDCs as Korea, Taiwan, Argentina, Brazil, and Hong Kong, and even industrially less advanced India. Whether these innovations have been worth the costs incurred is an issue raised by all the authors but is still unresolved. Because of the paucity of evidence and a presumption that costs exceed benefits, others have contended that LDCs should not pursue further technical independence and should be content to continue importing their technological needs. Stewart develops a number of arguments that can be adduced for LDCs to augment their technological capacity, including the fact that it is difficult for them to know which technologies to search for or the appropriate price to pay unless they command skills which are most likely to be generated from continued efforts at increasing their technological independence.

The various theoretical analyses and empirical findings point to a number of policy issues. An emphasis on measures to increase the international flow of technology at reduced costs to the LDCs, one of the items on the agenda of the New International Economic Order (NIEO), does not, in its simpler forms, have much to recommend it. There is likely to be a conflict, which McCulloch discusses, between increasing the short-run inflow of technology to LDCs especially via conditions favorable to them—low licensing fees—and the building of local capacity. Moreover, from the viewpoint of transferring firms, some of the policies may encourage increased short-term transfer at the cost of reduced longer-term transfer or vice versa, as Magee points out.

While most advocates of the NIEO would probably argue for import substitution policies to stimulate local industry, few seem to realize the same set of encouraging measures might be desirable vis-à-vis technology generation. However, it may be all too tempting to move too much in the direction of excessive protection or subsidization of technology development—indeed, such policies may be the last refuge of unrepentant import substitution proponents. While, as Stewart notes, a static calculus of benefits and costs of technology independence—or a move toward it—may lead to a premature conclusion that it should not be pursued very far, arguments about its long-run benefits in improving search and bargaining abilities may, given exiguous circumstances, be pushed too far, as was an earlier set of arguments about the long-range benefits of import substitution in commodities. It is difficult to derive precise policy guidelines that correctly balance the marginal benefits and costs accruing from a move along the path from total dependence toward some degree of technological independence. The benefits and costs will differ according to each country's previous actions and the resultant current endowment.

There is skepticism among several authors, for example, Magee, McCulloch, and Teece, that legal measures designed to reduce the price of acquiring technology or limiting the restrictions placed upon recipients—requiring the purchase of spare parts from the initial vendor—will improve the lot of most LDCs. The transferring firms may not realize sufficient profits to make the transfer effort worthwhile unless some of the restrictive practices remain, the profits being a substitute for still higher royalty payments. Furthermore, any international code along these lines that was successful would help most those who have already undertaken considerable domestic effort. Free access to all the proprietary knowledge in Silicon Valley will not help Chad or Nepal but would be of enormous benefit to Brazil or South Korea. In contrast, international codes designed to redress market imperfections, such as the costliness of information and the lack of competition among sellers, would largely aid those countries or firms that have invested little in their own learning, as McCulloch notes. Nevertheless, such codes would be of no help in enabling purchasers to utilize effectively the new equipment or processes.

There is an interesting contrast, which Sternheimer discusses, in the interactions between two developed countries that have pursued quite different paths to the development of nonmilitary technological capacity. The USSR, which has devoted limited resources to the augmentation of such capacity, remains quite dependent on imports, while Japan, whose strategy has been to build an ability to assimilate knowledge developed in other countries, has been able to base a good deal of its industrial growth on the skills so accumulated. Japanese experience may constitute one model for some of the advanced LDCs in terms of building a fast copying rather than a frontier ability with respect to new technology. In addition, it is known that Japan was highly selective in choosing among foreign technologies and was a tough bargainer. However, these are acquired skills obtained from considerable investment, not a genetic endowment. In this as in othe dimensions it is all too easy to conclude that Japan should be emulated, but technology policy, like other aspects of Japan's growth story, cannot necessarily be implanted into societies that differ in many other dimensions.

Finally, much of the official as well as journalistic discussion about technology transfer obscures a fundamental point: the industrial technology most appropriate for the great majority of LDCs is freely available on the international market as are the services of consulting firms whose *raison d'être* is the improvement of productivity. It is not proprietary information or the lack of specialized machinery that prevents Tanzanian or Philippine or Colombian firms from producing low-cost, acceptable quality textile fabrics. If anything, too much advanced technology—both machines and entire processes—is readily available at subsidized rates from developed country suppliers. For the basic industries from food processing to textiles to furniture production, the major need is for more judicious selection from the existing range of alternatives and more effective use of equipment and manpower. There is little that can be done in the international sphere to substitute for domestic effort in building up the skilled labor force and organizations required for such technological mastery, though correct incentives are likely to be a necessary condition. As the articles in this volume amply illustrate, successful development in agriculture and industry cannot be decreed—it depends on an imaginative and pragmatic long-range effort in individual countries.

HOWARD PACK

The Meaning of Technological Mastery in Relation to Transfer of Technology

By CARL J. DAHLMAN and LARRY E. WESTPHAL

ABSTRACT: The acquisition of technological mastery—that is, of the ability to make effective use of technological knowledge—is critical to the achievement of self-sustaining development. Transfers of technology are substitutes for local mastery rather than sources of it. Consequently, the part played by transfers of technology in the process of development, while important, is nonetheless limited. This article considers the role of technology transfer with specific reference to industrial technology, and places it in the broader context of the relationship between the acquisition of technological mastery and the development of an efficiently functioning economy. Based on a review of what is known about technical change in industrial enterprises in less-developed economies and on a case study of one economy's experience, it demonstrates that indigenous effort to assimilate technological knowledge is of overriding importance in the achievement of technological mastery. Various types of technological mastery are distinguished together with the different categories of effort associated with their acquisition. The consequences of increased mastery are also discussed, together with the factors that determine when it is appropriate to rely on transfers. Finally, the authors suggest that further research is needed to determine how technological mastery ought to evolve in relation to industrial development.

Carl J. Dahlman received a Ph.D. in economics from Yale University in 1979. Prior to joining the World Bank in the same year, he spent two years in Brazil studying technological change in the steel industry.

Larry E. Westphal received a Ph.D. in economics from Harvard University in 1969. Before joining the World Bank in 1974, he served on the faculties of Northwestern and Princeton universities, and for a time as resident advisor to the Economic Planning Board of the Republic of Korea. He has written extensively on industrial policy and development, on investment analysis under increasing returns, and on empirical analysis of production relationships and technological choice.

NOTE: The views and interpretations expressed here are the authors' and should not be attributed to the World Bank, its affiliated organizations, or any individual acting on behalf of these organizations.

THE exploitation of technological knowledge is central to the development process. Less-developed economies typically obtain this knowledge from more advanced ones rather than by creating it themselves. This is to be expected, given the vast pool of foreign technological knowledge available to them for exploitation. Transfers of technology are one means of acquiring foreign technological knowledge and can consequently play an important part in the development process. Nevertheless, their nature is such that transfers of technology can represent no more than an initial step in the exploitation of available knowledge.

Following are some important definitions.

—Technological Knowledge: Information about physical processes which underlies and is given operational expression in technology.

—Technology: a collection of physical processes which transforms inputs into outputs, together with the social arrangements—that is, organizational modes and procedural methods—which structure the activities involved in carrying out these transformations.

—Technological Effort: the use of technological knowledge together with other resources to assimilate or adapt existing technology, and/or create new technology.

—Technological Mastery: operational command over technological knowledge, manifested in the ability to use this knowledge effectively and achieved by the application of technological effort.

—Interrelationship of the Terms: technological mastery is the effective use of technological knowledge through continuing technological effort to assimilate, adapt, and/or create technology.

Technology is the translation into practice of technological knowledge. When technology is acquired by transfer, however, this process of translation is undertaken by foreigners. Transfers of technology thus substitute for indigenous technological mastery and make it possible to acquire technology without indigenous technological effort. Recognition of this fact leads to a question of central importance to less-developed countries: to what extent can effective use be made of available knowledge without indigenous effort to master it? This is the question dealt with in this article.

In order to give the discussion a manageable focus, our primary concern is with technological effort and mastery as they relate to physical processes.[1] This is not to deny the fact that these physical processes are undertaken within a framework of social arrangements—organizational modes and procedural methods—that condition their operation.[2] Similarly, we deal only with industrial technology. But it should be understood that the relationship of technology transfers to the acquisition of technological mastery is the same in all important respects for all sectors.

The sections that follow treat the transfer of technology in the context of a broader evaluation of the relationship between the acquisition of technological mastery and the development of an efficient indus-

1. Under this narrow definition, a firm or an economy could have a great deal of technological mastery and yet not deploy it effectively, owing to inappropriate organizational or procedural arrangements.
2. Harvey Brooks, "Technology, Evolution, and Purpose," *Daedalus*, 109 (1): 65-81 (Winter 1980); N. Bruce Hannay and Robert E. McGinn, "The Anatomy of Modern Technology: Prolegomenon to an Improved Public Policy for the Social Management of Technology," *Daedalus*, 109 (1): 25-53 (Winter 1980).

trial sector. The first section discusses what is meant by technological mastery and considers how it is related to transfer of technology and to technological effort. The empirical evidence for local technological effort and experience as sources of increased mastery and of the associated gains in industrial productivity is then summarized in the second section. Based on the perspective established in the preceding discussion, the third section outlines the factors that determine when it may nevertheless be appropriate to rely on transfers of technology rather than indigenous technological effort and indicates the various forms that transfers can take.

The evolution of technological mastery in relation to one country's industrial development is reviewed in section four. The case study is of the Republic of Korea, which has been chosen because of our comparative ignorance of other economies and—more important—because of the interest that attaches to understanding the sources of its rapid and highly successful industrialization. Finally, the concluding section highlights several important issues that have not yet received adequate attention in empirical research. These issues concern the relative efficacy of the alternative technological strategies that can be followed by less-developed economies.

MASTERY RESULTS FROM EFFORT, NOT TRANSFER

Industrial technology is sometimes misunderstood as being thoroughly documented in codified form—in "blueprints," as one prevalent metaphor would have it. If this simplistic view were valid, technologies could be transferred and assimilated effortlessly. Available evidence, however, belies this view in that ostensibly identical technologies are employed with vastly unequal levels of technical efficiency, or productivity, in different economies and even by different firms within a particluar one.[3]

Capital goods can be transferred, but capital goods alone do not constitute a technology; they represent only that part of the technology embodied in hardware. The remainder is comprised of disembodied knowledge—and although knowledge can be transferred, the ability to make effective use of it cannot be. This ability can only be acquired through indigenous technological effort, leading to technological mastery through human capital formation.

The application of technological knowledge within industry can usefully be broken down into four broadly defined categories of activities. In the order in which mastery is typically thought to be achieved in the development of particular industrial processes, they are as follows:

—production engineering, which relates to the operation of existing plants;

—project execution, which pertains to the establishment of new production capacity;

—capital goods manufacture, which consists of the embodiment of technological knowledge in physical facilities and equipment; and

—research and development (R&D), which consists of specialized activity to generate new technological knowledge.

3. Harvey Leibenstein, "Allocative Efficiency vs. 'X-Efficiency'," *American Economic Review*, 56: 392-415 (1966).

More will be said later about the acquisition of mastery in these activities. Several general observations are nonetheless in order at this point.

In the process of undertaking the first three activities, those carrying them out often find themselves involved in the solution of technical problems not previously encountered. Such problem solving represents an exercise of technological effort—that is, the use of technological knowledge to adapt technology—and may lead to a higher level of technological mastery. More generally, technological effort is also used in the assimilation or generation of new technological knowledge and hence in the invention of new technologies, which may be either adaptations of known technologies or radically new ones. Seen in this light, R&D is merely an extreme case, with respect to its degree of specialization, of the acquisition of new technological knowledge.

Technological mastery is a relative concept. Thus the extent of a firm's or an economy's mastery can be gauged only in relation to that of other entities. Moreover, mastery is not something that can be fully quantified. For one thing, it is possible to make unambiguous measurements of comparative technical efficiency only between entities that use ostensibly identical technologies. But—as we hope to make clear—technological mastery, even narrowly defined, involves far more than technical efficiency as conventionally understood. For example, an important aspect of mastery is the ability to adapt technologies so as to make them better suited to local circumstances—either by altering output characteristics to reflect local needs and preferences or by modifying input specifications

to permit the use of locally available materials and resources.

In addition, even if an entity's overall level of mastery could be measured, the separate contributions of the various types of mastery—corresponding to the categories of activity listed above—cannot be, because it is difficult to be precise about the interrelationships between them. This is particularly unfortunate, because many of the questions about technological mastery concern the relative importance of different types of mastery. For example, up to what point in a particular industry's development is mastery of production engineering sufficient? What is the relationship between mastery in production engineering and mastery in project execution? Is local capacity in capital goods manufacturing, or in R&D, necessary before socially warranted adaptations of technology can be made? These and similar questions can all be subsumed under a more general one: how should technological mastery in its various manifestations evolve in relation to industrial development? In addition to this question, the ensuing discussion deals with the question of how technological mastery is acquired.

EXPERIENCE AS A STIMULUS TO EFFORT

Technological mastery is not achieved by passively importing foreign technology. The extent of indigenous effort required for the successful assimilation of technology is most clearly demonstrated by case studies of technological changes that have occurred over time in individual firms. Much of this research has been prompted by dissatisfaction with a simplistic view of technology, which excludes the possibility that indigenous effort directed toward technological change in less-developed economies is an important part of the industrialization process.

The simplistic view holds that technology is something absolute and static: knowledge of a particular production technology either exists or it does not. A more realistic perception is that "manufacturing technology is characterized by a considerable element of tacitness, difficulties in imitation and teaching, and uncertainty regarding what modifications will work and what will not."[4] In other words, important elements of the technology appropriate to a particular situation can be acquired only through efforts to adapt existing technological knowledge. Any venture—for instance, the initiation of a new production activity—requires a great deal of iterative problem solving and experimentation as the original concept is refined and given practical expression. This sequential process lasts as long as changes continue to be made in the operation of the venture. Research on technological change at the firm level has demonstrated that this process can continue indefinitely, that it can produce technological changes that greatly increase productivity, and that it can yield substantially increased technological mastery.

Case studies of technological effort

Dahlman and Fonseca, for example, examined the technological history of an integrated Brazilian steel producer whose first plant was established with the help of Japanese steel makers.[5] In order subsequently to increase the plant's annual production capacity, the firm gradually built up its technological mastery through a carefully managed process of selectively importing technical assistance where needed to supplement its own engineering efforts. As a result, the plant's capacity was more than doubled from its initial nominal rating by means of a sequence of capacity-stretching technological changes implemented over seven years. Because these changes required very little additional capital investment and no additions to the work force, they more than doubled the plant's productivity. Moreover, as a result of the increased technological mastery this process stimulated, the firm was subsequently able to design and execute further additions to its capacity and to sell technical assistance to other steel producers, principally in Brazil, but elsewhere in Latin America as well.

More generally, firms in less-developed economies have been found to undertake substantial technological efforts in order to achieve a wide variety of technological changes.[6] These changes include, for

4. Richard R. Nelson "Innovation and Economic Development: Theoretical Retrospect and Prospect," IDB/ECLA/UNDP/IDRC Regional Program of Studies on Scientific and Technical Development in Latin America Working Paper, No. 31 (Buenos Aires: Economic Commission for Latin America, 1979), p. 18.

5. Carl J. Dahlman and Fernando Valadares Fonseca, "From Technological Dependence to Technological Development: The Case of the Usiminas Steel Plant in Brazil," IDB/ECLA/UNDP/IDRC Regional Program of Studies on Scientific and Technical Development in Latin America Working Paper, No. 21 (Buenos Aires: Economic Commission for Latin America, 1978).

6. The largest block of case-study research has been carried out under the auspices of the Regional Program of Studies on Scientific and Technical Development in Latin America, jointly sponsored by the Inter-American Development Bank, the United Nations Economic Commission for Latin America, the United Nations Development Program, and the International Development Research Center in Canada, and under

example, stretching the capacity of existing plants through various adaptations, as in the case just cited, breaking bottlenecks in particular processes within existing plants, improving the use of byproducts, extending the life of equipment, adjusting to changes in raw material sources, and altering the product mix. Some of the firms studied appear to have followed explicit technological strategies aimed at specific long-term objectives. Others seem merely to have reacted defensively to changes in their circumstances or to obvious needs to adapt imported technology. On the other hand, some firms have undertaken no appreciable technological effort and have consequently experienced no technological change.[7]

*Significance of
    indigenous effort*

Most of the technological changes uncovered in existing research can be characterized as minor, in the sense that they do not create radically new technologies, but rather adapt existing ones. Nonetheless, as shown by the example of the Brazil-

ian steel plant, a sequence of minor technological changes can have a pronounced cumulative effect on productivity. In fact, the cumulative sequence of technological changes following the initiation of a new activity may have a greater impact on the productivity of employed resources than that produced by its initial establishment.[8] This possibility has not, to our knowledge, been explored, but it is consistent with what has been learned about the process of technological change in the industrialized countries.

Studies of major technological changes in developed countries have found it useful to distinguish between what Enos refers to as the alpha and beta stages.[9] The former includes all efforts leading to and including the introduction of a radically new technology. The latter covers all of the subsequent minor technological changes undertaken to modify and adapt it. In his own analysis of the development and diffusion of six new petrochemical processes between 1913 and 1943, Enos found that the cumulative reduction achieved in production cost per unit during the beta stage was greater than the initial reduction obtained in the alpha stage. Studies show that other major technological changes have followed the same pattern.

From the standpoint of a developing economy, the assimilation of a technology newly imported from abroad is a major technological change. The initial transfer is parallel to Enos's alpha stage. The comparable beta stage is the subsequent,

the direction of Jorge Katz. For a summary of the research so far, see Jorge Katz, "Technological Change, Economic Development and Intra and Extra Regional Relations in Latin America," IDB/ECLA/UNDP/IDRC Regional Program of Studies on Scientific and Technical Development in Latin America Working Paper, No. 30 (Buenos Aires: Economic Commission for Latin America, 1978).

7. Martin Bell, Don Scott-Kemmis, and Wit Satyarakwit, "Learning and Technical Change in the Development of Manufacturing Industry: A Case Study of a Permanently Infant Enterprise," Science Policy Research Unit Working Paper (Brighton, Great Britain: University of Sussex, 1980); Ruth Pearson, "The Mexican Cement Industry: Technology, Market Structure and Growth," IDB/ECLA/UNDP/IDRC Regional Program of Studies on Scientific and Technical Development in Latin America Working Paper, No. 11 (Buenos Aires: Economic Commission for Latin America, 1977).

8. The reference here is to technological changes that occur after the achievement of predetermined project-specific norms—for example, the nominal capacity rating.

9. John L. Enos, "Invention and Innovation in the Petroleum Refining Industry," in *The Rate and Direction of Inventive Activity: Economic and Social Factors*, ed. Richard R. Nelson (Princeton: Princeton University Press, for the National Bureau of Economic Research, 1962), pp. 299-321.

gradual improvement in the productivity with which the technology is used. The relative significance of the beta stage for a developing economy's assimilation of a new technology appears to be much greater than the analogy suggests, however. To introduce a radically new technology into the world—as in Enos's alpha stage—requires mastery of that technology; by contrast, to import a technology —as in the technology transfer analogy—does not require mastery of it, at least not at the outset. Rather, the case study research suggests that it is in the beta stage that most of the increase in developing economies' technological mastery is achieved.

Only part of the impact of this increase is reflected in higher productivity using that particular technology; much of the impact spills over into related activities. For example, the mastery gained in assimilating one technology enables greater indigenous participation in subsequent transfers of related technologies, thereby increasing the effectiveness with which they are assimilated. A number of semiindustrial economies have even exploited their mastery to export technologies on a continually expanding scale to other developing economies.[10] In more general terms, the increased mastery that results from experience with previously established technologies contributes to an economy's capacity to undertake independent technological efforts, including replication or adaptation of foreign technologies as well as creation of new technologies.

Types of mastery acquired

Most of the technological changes so far uncovered can also be charac-

terized as having been derived from plant operating experience. Even within the confines of an existing plant, production processes do not remain static, certainly not if the firm is able to prosper within a relatively competitive environment. Production experience provides insight into how the operation of a plant can be altered to improve its performance. In addition, circumstances vary constantly over the life of a plant: input prices change, demand patterns shift, new competitors emerge, and so on.

This process of capitalizing on experience and reacting to varying circumstances requires continued technological effort to modify existing processes, which in turn represents an important source of increased mastery in production engineering—the first category of technological activity distinguished in the first section. Moreover, this form of technological effort often extends to changing the basic design of a plant, as when capacity is stretched or particular bottlenecks are broken. Thus it can also be a source of mastery in project execution—the second category in the typology provided in the first section. Nevertheless, although the type of technological mastery acquired through plant operating experience may overlap somewhat with that exemplified in project execution, the overlap can never be complete.

Mastery of almost all the tasks involved in project execution (see Table 1 for stages of project execution) requires extensive "learning by doing." Only for preinvestment feasibility studies does formal education alone suffice to impart the skills required. For the other tasks, the attainment of tech-

10. See Sanjaya Lall, "Indian Technology Exports and Technological Development," in

TABLE 1

STAGES OF PROJECT EXECUTION

1. *Preinvestment Feasibility Studies*, technical and economic, using readily available information to ascertain the viability of a project by examining alternative product mixes, input sources and specifications, plant scales and locations, and choices of production technology

2. *Detailed Studies*, following establishment of viability, using more specific engineering norms obtained from prospective sources of technology, leading to tentative choices among the alternatives considered previously and to refined estimates of capital requirements, personnel needs, cost and mode of financing, construction timetable, and the like

3. *Basic Engineering*, following confirmation of viability, to supply the core process technology by establishing the process flow through the plant and the associated material and energy balances, as well as designing specifications and layouts for major items of equipment and machinery

4. *Detailed Engineering*, to supply the peripheral technology, by providing complete specifications of equipment and materials, detailed architectural and civil engineering plans, construction specifications, installation specifications for all equipment, and the like

5. *Procurement*, which includes the choice of equipment suppliers and firms to construct and assemble the plant, coordination and control of the various subcontractors' activities and inspection of work in progress

6. *Training* of the plant's prospective personnel at all levels in various aspects of the plant's operation and maintenance, often through experience gained by working temporarily in a similar plant elsewhere

7. *Construction and Assembly* of the plant

8. *Startup* of operation, to attain predetermined project-specific norms and to complete the provision of training in the plant's operation

9. *Trouble-Shooting*, to overcome the various design problems encountered during the early years of the project's life

nological mastery requires previous experience in the same or closely related activities. Basic engineering, for example, calls for highly specialized knowledge of the core processes, which can frequently be gotten only through applied R&D, including pilot plant experimentation. Startup of operation often demands less familiarity with the principles underlying the core processes, but entails knowledge that can come only from previous production engineering experience in operating similar plants. Post-startup troubleshooting calls for somewhat more knowledge of the principles, but not necessarily as much as is involved in basic engineering. Detailed studies—the second stage of project execution—do not demand precise knowledge of individual core processes but do call for rather sophisticated knowledge of the industry. In turn, many of the individual detailed engineering tasks—for example, providing architectural and civil engineering plans that conform to requirements determined in the basic engineering stage—require no specialized knowledge whatsoever of the particular industry, but instead require other forms of specialized knowledge such as ability to design structures and civil works.

Production engineering and project execution are not the only broadly defined uses of technological knowledge, or types of technological mastery. Although they are not well incorporated into the existing research on technological change in developing countries, the two other categories of activity distinguished in the first section should not be overlooked. One is capital goods manufacture, which consists of embodying technology in machines. The other is specialized R&D to develop new products or processes.

These activities have strong links to production engineering and project execution, because to some degree they are prompted and given direction by the problems and opportunities that arise in connection with production and investment. Indeed, the kinds of technological effort associated with production engineering and project execution are frequently indistinguishable in concept from those involved in R&D. Likewise, these efforts often involve changes in the design of capital goods. Relatively little is known, however, about capital goods producers and specialized R&D performers as initiators of technological change, or about their roles in successful industrialization.[11]

RELIANCE ON TRANSFERS OF TECHNOLOGY

There are many means whereby less-developed economies can have access to foreign technological knowledge. Among them are various activities in which foreigners play a passive role, with the subsequent translation of this knowledge into technology being done indigenously. These activities include sending nationals abroad for education, training, and work experience; consulting technical and other journals; and copying foreign products. As Korean experience indicates—see the following section—these kinds of activities are tremendously important channels of information; almost invariably, some of the technological knowledge underlying new industrial initiatives in developing

countries comes via one or the other of them. By contrast, transfers of technology constitute a crucially different class of activities, in that the translation of technological knowledge into operational form is made by foreigners.

Whether technology should be obtained locally or from abroad should depend on the relative costs and benefits to the recipient of acquiring it from different sources. In this connection, the degree of local mastery in the various uses of the underlying technological knowledge is of critical importance. If little previous effort has been made to acquire mastery of the specific technology, reliance on domestic sources will entail either the replication—and perhaps also the adaptation—of foreign technology or the creation of new technology through indigenous effort. Local development, however, is rarely the most effective way of initially obtaining all of the necessary elements of a technology. More generally, an economy's capacity to provide the various elements depends on the stage of development of the relevant sector and those closely related to it.

Firms starting up or already engaged in traditional or well-established activities may often be able to acquire additional elements of technology relatively easily—either through their own developmental efforts or through the diffusion of expertise from other domestic firms. The hiring of personnel with previous work experience elsewhere plays an extremely important part in the diffusion of expertise among firms, as does the interchange of information among suppliers and users of individual products, especially in the case of intermediate products and capital goods. Firms engaged in newly or recently initiated activities typically have much less opportunity to take advantage of previous ex-

11. For surveys of what is known, see Howard Pack, "Fostering the Capital Goods Sector in LDCs: A Survey of Evidence and Requirements," World Bank Staff Working Paper, No. 376 (Washington, DC: The World Bank, 1980); Diana Crane, "Technological Innovation in Developing Countries: A Review of the Literature," *Research Policy*, 6: 374-95 (1977).

perience—if any—or of diffusion or explicit transfers from other domestic firms.[12] Firms in such a position are likely to find it more cost-effective to rely heavily on foreign suppliers of technology. Even in relatively highly developed sectors, selective transfers from abroad may be equally cost-effective as aids in the process of increasing productivity.

Modes of transfer

Transfers of technology take place in a large number of ways and often incorporate not only the translation of technological knowledge into information about operational processes but other elements as well. Imports of machinery—an extremely important mode of technology transfer—represent a case in point, in which the additional element is the embodiment of the technology in hardware. Another example is direct foreign investment when used as a means to acquire technology, with the additional elements typically being financial capital, management, and marketing.

Many modes of transfer do not involve explicit and separate payment for the transfer. This is frequently the case in the kinds of transactions instanced previously that incorporate additional elements, as it is with indirect technology transfers. As an example of the latter, exporting firms often receive valuable free technical assistance as a result of their dealings with foreign buyers; in the conduct of their normal business operations, these buyers frequently provide various forms of assistance in such areas as the upgrading of product specifica-

12. The opportunity is least when new process technologies must be mastered. It is much greater if the new activity simply involves applying known process technologies to the production of a new product.

tions and the achievement of improved quality control—see the following section outlining Korea's experience.

Explicit transactions to transfer technology without any other elements also take many forms. Among the simplest forms of transaction are contracts for the services of individuals or consulting companies to provide individual elements of technology—for example, to undertake specific design or process engineering tasks, to give technical assistance during various phases of the establishment and operation of a plant, or to provide technical information services. Other transactions include licensing and trademark agreements that transfer particular proprietary product and process designs.

The most all-inclusive form of transaction is a turnkey contract under which a general contractor is hired to assume complete responsibility for project execution, with the obligation to deliver an operating plant. Turnkey contracts, together with their counterpart in the form of direct foreign investment, are perhaps the most frequent mode of transferring technology for activities that are entirely new to an economy.

Turnkey contracts often deliver a plant together with instructions for operating it under the conditions assumed in its design, but they may fail to provide the recipient with an understanding of the full details of how the plant operates or of why it operates as it does. This hampers the recipient entity's ability to improve plant operating productivity or to adapt to changes that may occur over time in the circumstances that affect how the plant is best operated. As a result, the plant is likely to operate at lower productivity than could optimally have been achieved, with the entity probably also continuing to depend excessively on foreign mastery for technical assistance in trouble-shooting. Alternatively, the entity will need to make greater efforts to achieve internal mastery than would have been needed if more complete information had initially been provided. These outcomes can be avoided by having the entity's personnel participate in every phase of project execution, even if only as intelligent observers who merely follow the work in progress and learn which are the relevant questions in gaining mastery of the "hows" and "whys."

The foregoing discussion points to the possibility that government intervention might be warranted to ensure that transfers of technology contribute appropriately to the development of indigenous technological mastery. This possibility raises a variety of issues, many of which are dealt with in the other articles in this issue. For the purpose of the present discussion, however, it is relevant to examine one important component of the knowledge required to design effective policies—the question of how technological mastery should evolve in relation to industrial development. It is to this issue that we now turn.

KOREAN TECHNOLOGICAL MASTERY

Historical evidence forms the principal basis for considering the relationship of technological mastery to industrial development. The Republic of Korea—often referred to as South Korea and hereafter simply as Korea—provides an instructive example. The broad outlines of Korea's remarkably successful achievement of semiindustrial status are well known and need not be repeated here. Less well known are what Korea's technological mastery consists of and how it was acquired. Available evidence on these points is summarized below for the period from the end of the Korean war through approximately 1978.[13]

The fundamental elements of Korea's industrialization have been directed and controlled by nationals. Foreign resources have made substantial contributions, but the transactions involved have typically been at arm's length. Thus, although Korea has relied quite heavily on capital inflows, these have overwhelmingly been in the form of debt, not equity, and technology has been acquired from abroad largely through means other than direct foreign investment. The purchase of technology through licensing agreements has been of modest importance as the initial source of process technology. Machinery imports and turnkey contracts have been of much greater consequence in the transfer of technology, and a tremendous amount of expertise has been obtained as a result of the return of Koreans from study or work abroad. Moreover, in only a few sectors—such as electronics—have Korean exports depended critically on transactions between related affiliates of multinational corporations or on international subcontracting.[14]

Nature of the technologies mastered

Korea's success in assimilating technologies acquired through

13. The following discussion is based on detailed evidence given in Larry E. Westphal, Yung W. Rhee, and Garry Pursell, "Foreign Influences on Korean Industrial Development," *Oxford Bulletin of Economics and Statistics*, 41: 359-88 (1979).

14. International subcontracting refers to

arm's-length transactions is in part explained by the nature of technology and product differentiation in the industries on which its growth has crucially depended. Many of these industries—such as plywood or textiles and apparel—use relatively mature technologies; in such cases, mastery of well-established and conventional methods, embodied in equipment readily available from foreign suppliers, is sufficient to permit efficient production.[15] The products of many of these industries are either quite highly standardized, plywood, for example, or differentiated in technologically minor respects and not greatly dependent on brand recognition for purchaser acceptance, for example, textiles and apparel. Thus, in most of the industries that have been intensively developed, few advantages are to be gained from licensing or direct foreign investment as far as technology acquisition and overseas marketing are concerned.

Nonetheless, exceptions exist, of which electronics is perhaps the most notable. This is an industry in which technology is changing rapidly worldwide, product differentiation is based on sophisticated technological expertise, and purchasers' brand preferences are evident. Given these characteristics, it is not surprising to find that in this case Korea has relied extensively on direct foreign investment to establish production, particularly for

export, and has so far failed to gain local mastery of many key aspects of production engineering. It should be noted, however, that the electronics and certain chemicals industries are unique in Korea in their almost exclusive reliance on direct foreign investment for acquiring the very latest technology and market access.

In other industries, where technology is similarly proprietary, a number of examples attest to the fact that Korean industry has managed to initiate—and in most cases to operate successfully—a variety of "high technology" industrial activities by means of licensing and turnkey arrangements. To cite two cases: Korea used arrangements of this kind to acquire the most modern shipbuilding technology in the world and to incorporate the most recent technological advances in its integrated steel mill. More generally, Korea's recent experience in promoting technologically sophisticated industries indicates that their development may involve greater reliance on licensing as a way of acquiring technology.

Activities leading to mastery

Korea's past strategy for gaining technological mastery has relied heavily on indigenous effort through capitalizing on experience and emphasizing the selective use of technology transfers. In industries for which process technology is not product-specific, the initial achievement of mastery has frequently permitted the copying of foreign products as a means of enlarging technological capacity. The mechanical engineering industries, among others, afford many examples; such processes as machining and casting, once learned by producing one item, can readily be applied

---

export activity that is wholly organized by an overseas firm; the domestic, exporting firm is responsible only for overseeing production.

15. This does not imply the absence of rapid technological change in the industry in developed countries. It simply means that developing countries can—at least for a while—maintain a comparative advantage, once established, based on mastery of conventional methods more appropriate to their factor endowments.

in the production of others. One case that has been closely studied is textile machinery, in particular semiautomatic looms for weaving fabric.[16] In this as in some other cases, Korean manufacturers not only have been able to produce a capital good that meets world standards, albeit of an older vintage, but have, in addition, adapted the product design to make it more appropriate to Korean circumstances; the adapted semiautomatic looms fall between ordinary semiautomatic and fully automatic looms in terms of the labor intensity of the weaving technology they embody. In other industries in which technology is more product-specific, such as chemicals, mastery of the underlying principles has permitted greater local participation in the subsequent establishment of closely allied lines of production.

Export activity has proved to be a very important means of acquiring technological mastery. As a result of exporting, Korean firms have enjoyed virtually costless access to a tremendous range of information, diffused to them in various ways by the buyers of their exports. The resulting minor technological changes have significantly increased production efficiency, changed product designs, upgraded quality, and improved management practices. Exporting thus appears to have offered a direct means of improving productivity, in addition to the indirect stimulus derived from trying to maintain and increase penetration in overseas markets. The Korean experience also suggests that this beneficial externality of export activity may partly explain why

countries following an export-led strategy have experienced such remarkable success in their industrialization efforts.

In addition, the fast pace of Korea's industrial growth has permitted rapid rates of technological learning because of the short intervals between the construction of successive plants in many industries. In some industries, including synthetic resins and fibers, the first plants were often built on a turnkey basis and on a scale much smaller than either that warranted by the size of the market or that which would exhaust economies of scale. Construction of the second and subsequent plants—at scales much closer or equal to world scale—followed quickly, with Korean engineers and technicians assuming a gradually increasing role in project execution.[17]

Significance of the Korean experience

Korea's experience demonstrates that a high level of technological mastery in all aspects of the uses of technological knowledge is not required for sustained industrial development. This is evident from the fact that its mastery has progressed much further in production engineering than in project execution. In addition, Korea has relied on foreign suppliers for necessary capital equipment and has only recently embarked on a concerted program of import substitution in

16. Yung W. Rhee and Larry E. Westphal, "A Micro, Econometric Investigation of Choice of Technology," *Journal of Development Economics*, 4: 205-38 (1977).

17. The observed pattern of time-phased plant construction in these industries might be an optimal strategy, with small scales chosen for the first plants to minimize the costs and risks entailed in learning the technology. It is not known, however, whether these or other considerations were the controlling ones at the time the first plants were constructed.

the capital goods sector. Nonetheless, Korean industry has acquired and exercised the capacity to choose the technologies to be imported, and Koreans have become increasingly involved in other phases of project execution. Fundamentally, however, Korea has become a significant industrial power mainly as a result of its proficiency in production. It thus appears that mastery of production engineering alone is nearly sufficient for the attainment of an advanced stage of industrial development.

Contemporary pronouncements about the nature of, and the constraints imposed by, the existing international economic order are contradicted by Korea's experience. In the context of calls for a "new international economic order," it is frequently alleged that existing international markets are noncompetitive and that developing countries are either denied access to technology and overseas markets or are granted it only on highly unfavorable terms. It is further asserted that foreigners exercise the initiative in transfers of technology and in the organization of export activity. If true, these assertions would imply a severe constraint on industrial development. Far from supporting them, Korea's experience shows them to be false for many important industries.

To summarize, in the course of its industrialization, Korea has effectively assimilated various elements of foreign technology. Transfers of technology have contributed importantly to this process. A wide variety of transfer modes has been used, with machinery imports and turnkey contracts predominating over licensing agreements and direct foreign investment in the initial acquisition of technology. But transfers have been no more than an initial step in the exploitation of available knowledge. Assimilation has been achieved through a succession of technological efforts over time, largely undertaken by domestic firms to extend their technological mastery and to accomplish minor technological changes. These efforts have resulted in continual and significant increases in the productivity of resources employed in the industrial sector and have been reflected in Korea's sustained rapid industrial growth. Korea's experience thus supports the argument that indigenous effort is of overriding importance in the achievement of technological mastery, but the causal forces that contribute both to the presence and to the effectiveness of indigenous effort have yet to be uncovered.

CONCLUSION: ISSUES OF TECHNOLOGICAL STRATEGY

The dependence of an economy's fund of technological expertise on the mastery of previously introduced technologies has important implications. It means that initial decisions about choices of technology and degrees of local involvement in investments to implement them are critical determinants of the directions in which an economy's technological mastery will develop. Although the empirical evidence derived from research is not yet comprehensive enough to provide a clear basis on which to make prescriptions about how an economy's technological mastery ought to evolve in relation to its industrial development, it seems clear that a synergistic relationship can develop between them, with advances in each prompting new gains in the other. As Korean experience demonstrates, however, high indigenous levels of all types of technological mastery are not necessary for the initial stages of industrial development; in the Korean case, a mastery that has been mainly confined to production engineering has been sufficient. The Korean example also suggests that by relying on foreign sources of technology, it is possible to choose a technology without having first mastered its use. In the same way, it is also possible to use a technology without having the mastery required to replicate it through project execution or to manufacture the capital goods involved.

Nevertheless, it should be remembered that, just as the initial choice of production method may greatly constrain the direction of technical change, so the kinds of technological effort in which an economy acquires experience may constrain the type of technological mastery it can develop. Furthermore, there is an important difference between attaining mastery in relation to given circumstances and in attaining the capacity to adapt to changing circumstances. The objective of acquiring technological mastery is not simply to produce in the present; equally, it is the ability to adapt technology and to anticipate changes in world and domestic markets. Thus it is also necessary to develop the capacity to innovate in various respects. It is unclear how far this capacity can be developed solely on the basis of production engineering or project execution experience.

The effects of government policy on the development of indigenous technological mastery have yet to be ascertained. Further research to uncover historical evidence from different countries' cases is necessary to reach any soundly based generalizations about the determinants of the extent and appropriateness of technological effort in different directions. Such generalizations are needed to formulate policies that will direct the attainment of increased technological mastery in ways in line with social objectives. In particular, much remains to be learned about the appropriate phasing of the replacement of technology transfers by indigenous technological effort and about the impact of different policies on the development of the various types of technological mastery.

ANNALS, *AAPSS*, 458, November 1981

Appropriate Industrial Technology:
Benefits and Obstacles

By HOWARD PACK

ABSTRACT: The systematic adoption of appropriate rather than advanced industrial technology in the modern manufacturing sector of less-developed countries could increase output and employment by substantial amounts. Calculations of the rough order of magnitude of these gains are presented. It is argued that among the factors inhibiting the realization of these benefits are large information costs incurred in learning about technical options as well as the difficulties governments perceive in altering current economic policies that discourage the use of appropriate technology.

Howard Pack received a Bachelor's degree in economics from the City College of New York and a Ph.D. in economics from M.I.T. He has taught at Yale University and is currently professor of economics at Swarthmore College. He has written about many issues arising during the process of industrialization, including the choice of appropriate technology, the role of the capital goods industry, and the determinants of industrial efficiency.

THE eyes of readers of the *New York Times* or the *Wall Street Journal* are likely to become glazed when confronted yet again with the expression "appropriate technology," currently among the most chic if not the pithiest of phrases. If buzzwords could be copyrighted or become trademarks, a fierce battle for proprietary rights would be staged by all manner of interest groups, including advocates of windmills, solar power, log cabins in Vermont, and importers of pyrethrum. For economists, forever mired in their dismal science, concern with appropriate technology has arisen largely in connection with the industrialization policies of less-developed countries (LDCs). Unlike the futurologists of solar power, proponents of the correct choice of industrial technology, to use the full and awkward phrase, often look slightly backward in time and even occasionally to the East for a source of machinery if not of inspiration.

NATURE OF THE PROBLEM

A numerical illustration of the difficulty encountered in providing an adequate number of jobs for new entrants to the labor force can help us to understand the importance of appropriate technology in the context of less-developed countries. Consider an economy with a total population of 1000 and per capita income of $1000—countries such as Guatemala, Malaysia, and Tunisia are roughly at this level of income. If the country has a saving rate, net of the amount required to replace worn assets, of 15 percent, $150,000 will be available to satisfy all investment requirements. Assuming that $60,000 of this total needs to be allocated to "social" investment such as

housing, roads, and schools, $90,000 remains to purchase agricultural irrigation equipment, seeds, and industrial plant and equipment. If the labor force is 50 percent of total population and is growing at 3 percent per annum, it increases by 15 workers per year.

The $90,000 of available investment funds may be divided equally among the new labor force members, providing each with $6000 of capital. If, however, an industrial project is undertaken in which each of three new workers is endowed with $28,000 of plant and equipment, only $6000 will remain to provide the means of production to the 12 other new entrants.

The effects of this second scenario are easy to deduce, and the available evidence yields painful corroboration of these deductions. The workers employed in firms using large amounts of capital will be much more productive, in the sense that each will produce a higher product, though not because of any superiority in native or acquired skill. Conversely, those not lucky enough—and it is largely a matter of luck—to benefit from the largesse will be condemned to work with small amounts—$500—of machinery, irrigation piping, seeds, or inventories, and will earn commensurately small incomes. A small fraction of the labor force will earn incomes three or four times that of an equally skilled, often longer- and harder-working majority.

For the purpose of this article, the decision to begin a new factory that requires an investment of $28,000 per worker where only $6000 is available, on the average, is an inappropriate choice of technology, the appropriate choice being a technology that results in something approaching the latter figure. In general, the investment per job will be closely related to the sophistication of equipment, machines with more labor-saving devices and mechanisms that allow automatic transfer of work-in-process between operations being more costly. One qualification should be noted, namely, virtually identical machines made under license may cost less in, for example, Japan than in Germany or Switzerland. It is easy to demonstrate analytically, and examples will be given subsequently, that the economy-wide choice of appropriate technology will raise national income, total wages, and nonwage or profit income as compared with a systematic choice of advanced, more expensive plant and equipment.

A skeptical reader might ask at this point whether it is possible to identify products and their processes of production that require investment per worker of only $6000, a relatively low figure by the standards of the industrialized countries. While it will be seen shortly that such products and processes can indeed be found, the $6000 criterion—and a lower one in countries with less per capita income—may be too stringent. Most LDCs continue to be largely agricultural, and the per-worker investment necessary in this sector is much below $6000. On the average two- to five-acre farm, typical of many countries, a $3000 tractor is decidedly inappropriate. Thus the amount of investment per worker available for urban activities, including manufacturing, will generally exceed the economy-wide average but will inevitably fall much below the amount available in the industrialized countries. It is this divergence between the advanced and poorer countries that

leads to the appropriate technology issue—replication of the products and processes characteristic of the former will result in lower incomes and employment for the country as a whole.

The remainder of this article will be devoted to analyzing questions that arise in connection with the choice of industrial technology. Other articles in this volume consider the agricultural sector. The focus here is on production alternatives and their implications in the large-scale industrial sector rather than among small firms. Nevertheless, even in advanced economies many industrial products continue to be produced by small- and medium-scale enterprises. Extension of the analysis to these products and processes would lead to an even more optimistic picture of the potential benefits to be obtained from correct choices. However, less precise empirical evidence is available for those sectors; hence the concentration on large-scale manufacturing.

STAGES OF THOUGHT ABOUT APPROPRIATE TECHNOLOGY

In the years following World War II there was great optimism that the pool of accumulated industrial technology could be readily borrowed by poorer countries, thus avoiding the costly process of in-plant learning and explicit research and development that had generated the current worldwide stock of technology. The "benefits of backwardness" or of being a latecomer to industrialization gained a considerable audience in some universities and aid-giving agencies, and undoubtedly caught the attention of many persons then being trained to assume decision-making positions in countries in the process of attaining independence.

The appeal of newer technology was reinforced by an argument often heard, crystallized in an influential article in the mid-fifties, that newer technologies that used fewer workers generated a greater level of reinvestable profits that could be utilized to finance still further purchases of investment goods.[1] Even if employment were lower in the early stages of development, the greater reinvestment level would eventually create a larger capital stock and hence more jobs than would a policy of labor-intensive development from the beginning. Although this view was based on a strong set of untested— and debatable—assumptions about critical parameters, it quickly gained numerous converts and supplied a rationalization for policies that many governments clearly favored in any case. For a decade this view of the desirability of capital-intensive development, and the associated belief that trickle-down would be sufficient to help those not fortunate enough to benefit directly from the policy, constituted the implicit guideline followed in almost all LDCs. The creation of employment and the distribution of income were, at best, ignored.

Only in the late 1960s, when it had become evident that relatively few jobs were being created in the manufacturing sector of less-developed countries despite the rapid growth in production and that a serious maldistribution of income had begun to manifest itself, did the questions of appropriate technology —or the factor proportions problem, in the economist's lexicon—again

become the subject of intense investigation.

The research done in the ensuing decade and a half has been of sufficient volume to have elicited at least five major survey articles.[2] Although there are some dissenters, the view of most scholars working on these problems, and certainly the conclusion of the authors of the surveys, is that in many industrial activities it is possible to establish or expand a plant with a technology that uses more labor and less capital than would be utilized in advanced economies in Western Europe or North America. Much of the potential substitution of labor for capital stems from use of labor-intensive methods in "peripheral" production activities; labor, with little if any capital, can be used to transport material efficiently within the factory, to pack cartons, and to store the final product. The evidence for these statements is drawn from observation of both developed country (DC) and LDC factory operations and engineering specifications.

Evidence also exists that the core production process itself, whether

1. W. Galenson and H. Leibenstein, "Investment Criteria, Productivity, and Growth," *Quarterly Journal of Economics*, 69(3):343-70. (Aug. 1955).

2. The surveys are: S.N. Acharya, "Fiscal/Financial Intervention, Factor Prices, and Factor Proportions: A Review of the Issues," World Bank Staff Working Paper No. 183 (Washington, DC: World Bank, 1974); J. Guade, "Capital-Labor Substitution Possibilities: A Review of Empirical Evidence," in *Technology and Employment in Industry*, ed. A. Bhalla (Geneva: International Labour Office, 1975); D. Morawetz, "Employment Implications of Industrialization in Developing Countries—A Survey," *Economic Journal*, (335):491-542 (Sept. 1974); F. Stewart, "Technology and Employment in LDCs," in *Employment in Developing Nations*, ed. E. Edwards (New York: Columbia University Press, 1974); and L. White, "The Evidence on Appropriate Factor Proportions for Manufacturing in Less Developed Countries: A Survey," *Economic Development and Cultural Change*, 27(1):27-60 (Oct. 1978).

cooking of food or production of yarn, offers efficient possibilities for using less expensive equipment and more labor per unit of output. Adaptation of existing equipment—for example, changing the "normal" speed of operation—offers additional opportunities to save capital and increase the relative use of labor.

The extent to which labor can be substituted for capital varies across industrial sectors, being greater in food processing and some parts of textile manufacturing than in steel or fertilizer production. Whatever the appeal to national pride, steel, fertilizer, and other products whose production can be carried out only with very large amounts of capital per worker are inappropriate products for local manufacturing in most poor countries. The basic problem is to forestall the introduction of such plants, any minor choice of technology that does exist being of distinctly secondary importance. Indeed, a good part of the success of the rapidly growing East Asian countries has been their generally correct sectoral emphasis as well as the selection within sectors of appropriate technology. The fact that some of these countries—for example, Korea—are now attempting to upgrade their industrial structure to steel and nonelectrical machinery does not vitiate the general principle; this is being done after the success of the labor-intensive strategy and in response to one symptom of this success, namely, a growth in real wages resulting from the increasing scarcity of labor. Graduation to more capital- and technology-intensive sectors is the reward—and, given the difficulty, the punishment—for good performance in the early industrialization process.

BENEFITS OF APPROPRIATE TECHNOLOGICAL CHOICE

The following presents briefly some estimates of the benefits to be obtained by a typical poor country from carefully choosing appropriate technology rather than more advanced technology.[3] The benefits include increased income produced by the industrial sector, greater wage and nonwage—profit—levels, and greater employment. To establish some orders of magnitude it is necessary to specify both the goods to be manufactured and the alternative methods available for their production. The products and two values of investment per worker with which each can be manufactured are shown in Table 1. These goods are currently manufactured in many LDCs and, for better or worse, are high on the priority list when an expansion of industrial production is considered.

The products include some about which it is generally assumed that choice in production method is physically feasible—shoes, yarn, and woven cloth—as well as those about which conventional wisdom contends that not much variation is possible—fertilizers and beer. The figures in the advanced technology column indicate the amount of investment per worker that would be required if an LDC plant were to be established with the same core machinery and material transfer mechanisms as are used in a developed country. The second column shows the amount of investment when an appropriate technology is

3. The discussion in this section is drawn from my paper, "Macroeconomic Implications of Factor Substitution in Industrial Processes," World Bank Staff Working Paper No. 377 (Washington, DC: the World Bank, March 1980).

TABLE 1
BENEFITS FROM ADOPTING APPROPRIATE TECHNOLOGIES

| TECHNOLOGY | VALUE ADDED | WAGE INCOME | NONLABOR INCOME | EMPLOY-MENT | CAPITAL-LABOR RATIO | VALUE ADDED PER WORKER |
	(millions of $ a year)			(workers)	($ per worker)	($)
Appropriate	624	119	505	238,678	3771	2614
Advanced	364	29	335	58,017	15,513	6274

SOURCE: H. Pack, "Macroeconomic Implications of Factor Substitution in Industrial Processes," World Bank Staff Working Paper No. 377 (Washington, DC: the World Bank). Table 2.

purchased. Appropriate, here, is defined as the combination of labor and equipment that maximizes the profitability to the firm, whether private or state owned, and it can be shown that this definition conforms to the requirement that the amount of investment per worker not be excessive in the sense used earlier.

The appropriate technology shown is not necessarily the most labor-intensive production method available if those that require even smaller investment per worker are less profitable. With the exception of fertilizer, the difference between advanced and appropriate technology is very large. Equally impressive is the variation among products even when the choice is restricted only to appropriate technologies. The potential impact of selecting the right product as well as the correct process is clear.

Envision a country planning to establish new production capacity in each of the listed products, and for simplicity assume that $100 million is to be invested in one year in each of the sectors. What is the effect on national income, total wages, total profits, and employment of systematically choosing the appropriate rather than the advanced technol-

ogy? Rather than present the results sector by sector, Table 1 presents a summary of the impact for the nine products. The level of national income produced by the large-scale manufacturing sector can be increased by 71 percent, total wage payments by 311 percent, total profits by 51 percent, and employment by 311 percent. The significance of these figures can be stated in a number of ways. For example, for employment to be increased commensurately if advanced technology were chosen would require investment to be 300 percent greater, an impossible achievement in most countries. For the few who could manage it, an extremely painful suppression of the current standard of living would be required to obtain the requisite amount of saving. From another perspective, the effect of the proper choice of production method on national income produced by manufacturing is equivalent to 10 years of industrial growth at a rate of 5.5 percent per annum.

The impact of technology choice on disparities among individuals and households within a country can be inferred from the last three columns of Table 1. Employment in modern sector manufacturing plants is much greater when appropriate rather than capital-invensive plants are adopted. Thus a smaller percentage of the labor force is forced into marginal occupations, such as street vending or occasional artisan activity. Each worker employed in the large-scale sector is provided with a smaller amount of investment goods with which to work (column 5, capital-labor ratio) and hence produces less (column 6, value added per worker). Thus not only is a smaller percentage of the labor force engaged in low-productivity jobs, but the difference in income between those in the large-scale sector and those in marginal pursuits is decreased. Nevertheless, with the more even distribution of investment, total output is much greater and, as noted previously, the total amount of both wages and profits—or surplus for government enterprises—also increases.

Any politician in the United Kingdom or the United States faced with so seemingly simple a decision, for which a "yea" unleashes such a beneficial flow of results, widely distributed over all groups, might take one of two actions: quickly offer to exchange places with Mrs. Gandhi or Mr. Nyerere, or, staring very hard down the mouth of the gift horse, look for defective molars in the form of potential opposition or unacknowledged costs. And he or she would be well advised to take some time for a careful search, especially as there are usually unbooked seats to Delhi and Dar. Why, in practice, is the "simple" correct decision so rarely made? Are businessmen and bureaucrats misinformed, or is the implementation of the necessary policies too complex for the typical LDC?

SOME OBSTACLES TO CORRECT DECISIONS

Just as where one stands on a political issue often depends on the side of the legislative aisle on which one sits, the relative importance of the various obstacles to pursuing an appropriate industrialization policy, as seen by analysts of the problem, depends on more general perceptions about the process of economic development. Though almost all analysts note the same set of problems, the emphasis on one or another is often derived from attitudes toward public versus private ownership, the correct role for multinational as compared with domestic ownership, and the role of markets for both products and factors of production.

Type of ownership

Some scholars have argued that private owners of firms are anxious to avoid dealing with large numbers of workers. While this piece of casual empiricism is undoubtedly partly correct, it does not follow that public enterprises choose more appropriate technology. The limited amount of systematic evidence on this question suggests that both types of firms choose quite similar equipment in countries where both operate in the same sector—for example, Turkey. In countries such as Tanzania and Egypt, with few private firms, such comparisons are not possible; however, in these countries public sector firms appear to have chosen similar machinery to that employed in neighboring countries by privately owned companies.

A different binary classification is often thought to be of decisive importance, namely, the difference between foreign and domestically owned firms, regardless of whether the latter are public or private. The entirely a priori argument suggests that multinational corporations (MNCs) locating in poor countries will replicate the technology used in the home country of the enterprise, disregarding the needs of the LDC in pursuit of its own profits. Underlying this argument is the assumption that the cost of modifying a technology exceeds the saving in production cost that can be realized by using more labor—whose cost is lower—in the LDC than in the home country. While plausible, the issue can be resolved only by an appeal to systematic evidence, a substantial body of which has recently been accumulated.

Though individual anecdotes abound of highly automated plants being introduced by MNCs, several dozen studies using comparisons within countries of domestic and foreign firms producing the identical product largely support the hypothesis that no differences exist in the choice of plant and equipment.[4] Where differences are present, it is the MNCs that typically show more adaptation of technology. One of the most recent studies produces the typical result in height-

ened fashion. Donald Lecraw compared three types of companies in Thailand: ones that are domestically owned; MNCs whose home country is another LDC, such as India or Hong Kong; and MNCs from the developed countries.[5] The highest capital per worker was exhibited by the domestic firms; the lowest by LDC-based multinationals. These and other results do not demonstrate the benevolence of MNCs nor the malevolence of domestic firms. Rather, they are manifestations of determinants of technology choice more basic than nationality of ownership—for example, the cost of obtaining appropriate equipment. This issue will be discussed next.

Cost of labor and capital and competitive markets

An important argument noted by economists of all political views, though emphasized more by those who believe in the power of prices as a determinant of behavior, is the role of the cost of labor and equipment and the extent of competitive pressure. A company beginning or expanding its operations may adopt a variety of methods of production, the actual decision depending on the costs of the factors of production: labor, capital, and raw materials. It has become conventional to assert that labor costs are "too" high and capital costs too low in LDCs. What is the precise meaning of these statements? I briefly consider each of the two factor costs, wages and the cost of plant and equipment, in addition to competition.

4. See H. Pack, "Technology and Employment: Constraints on Optimal Performance," in *Technology and Economic Development: A Realistic Perspective*, ed. S. Rosenblatt (Boulder, CO: Westview Press, 1979) for a survey of the literature on the technology choices of multinational corporations.

5. D. Lecraw, "Direct Investment by Firms from Less Developed Countries," *Oxford Economic Papers*, 23(9):442-57 (Nov. 1977).

Hiring a worker entails the payment of a wage and one or more of the following: payments in kind—housing—fringe benefits—social security—and in some countries such supplements as "thirteenth month" salary. The cost of hiring a worker is "too" high if the value of the cash wage and other benefits exceeds the income the worker could command elsewhere, given his abilities, both inherited and obtained by education and on-the-job experience. It has long been noted that the typical employee in a modern enterprise, be it a factory, bank office, or government agency, earns considerably more than a worker in small-scale artisan shops or in self-employment, such as barbering. Modern employment also provides incomes considerably in excess of that of agricultural workers and small-scale peasant farmers. It is generally believed that the observed income differentials do not represent a reward for greater productive ability, but are artificially high and institutionally supported, reflecting government minimum wage legislation, union bargaining success, and a guilty aversion to paying lower, more appropriate wages that typify other activities.

The statement that wages are too high thus refers to the norm of alternate income possibilities for a similarly skilled worker, either in the urban craft sector or in a variety of rural activities. It does not imply that these wages are excessive in comparison with those in developed countries or that such workers are able to afford a luxurious living standard.

The cost of utilizing plant and equipment reflects the purchase price of a factory building or machine and the interests costs incurred in financing it. More precisely, the cost of using plant or equipment is best viewed in terms of the annual expenditure—depreciation of the initial acquisition cost and a yearly financing charge incurred as a result of a decision to purchase the capital item. The purchase cost of equipment is too low in most LDCs, in the sense that the net effect of government foreign trade policies is typically to lower artificially the amount of domestic currency that must be given up to pay for an imported machine. For many industrial products this artificial cheapening of foreign goods is offset by a relatively high, often prohibitive tariff imposed on imported goods that compete with domestically produced goods. However, no tariff is imposed on imported equipment, ostensibly to encourgae domestic investment. Thus LDC firms purchasing new equipment pay a lower price than they would if governments did not discriminate among different types of imported goods. A low purchase price is reflected in low annual depreciation charges, one of the two major components of the annual cost of using equipment.

As mentioned earlier, the second major cost is the financing charge. The interest rate paid by larger companies in the urban sector is too low as a result of governmentally imposed limitations on the rate of interest. At the existing low ceiling levels of rates, the total demand for funds exceeds the supply, and the existing supply is rationed among competing companies, none of which is charged more than the legal maximum. Companies that are unsuccessful in this competition are forced to compete in a grey or black market in which the rates often are three or four times the official one. The successful, usually large, firms thus in effect receive subsidized loans.

Apart from measures that lead to too low a purchase price and interest rate for many investors, numerous tax regulations further reduce the annual charge for using equipment. For example, investment credits and accelerated depreciation are likely to have adverse effects on the choice of production methods, particularly in view of the already high rates of return being earned by investors, who hardly require additional incentives.

The net effect of the existing set of distortions in wages and the cost of capital has presumably been to bias the choice of individual firms toward production methods that use unnecessarily complex machines rather than unskilled labor.

The ratio of labor to capital costs is usually assumed to play a decisive role in determining the relative amounts of capital and labor used in the production process. The importance of factor prices flows from the assumption of a competitive milieu. Factor prices play a more limited role, however, in noncompetitive environments. A firm currently realizing a 60 percent rate of return on equity capital, though using an inappropriately high ratio of capital to labor, may have little incentive to search for more appropriate methods that raise its return to 65 percent. The losses from forgone leisure and the difficulties often alleged to result from management of a larger labor force make such behavior perfectly plausible. If factor prices are to exert pressure toward adopting appropriate technology, some competitive forces must be present. Given the small markets typical of many LDCs, such pressures are best engendered by international competition rather

than by the proliferation of large numbers of small domestic companies, none of which is likely to reach economically efficient size. In the presence of high rates of tariff protection, changes in relative factor prices may have some beneficial effects, but these are likely to be highly attenuated. Thus an integral component of any determined effort to achieve more desirable factor proportions must be some increase in competitiveness in the product markets in which industrial firms participate.

Political economy of appropriate technology

Even if factor prices do affect the decisions of enterprises, altering them to obtain the potential gains shown in Table 1 is likely to require abilities that exceed the talents of all but a few of the leaders in any generation, for such alteration involves nothing more than revising the existing rules by which individuals earn income in a society. Union workers and those covered by effective minimum wage legislation would have to accept a decrease in their wages relative to the incomes of the marginal urban workers and the rural poor; firms receiving subsidized loans at a 10 percent annual rate of interest would have to pay 30 or 40 percent real—after inflation—rates of interest and would be subjected to increased competitive pressure as protective tariffs were reduced. Bureaucrats running the complex controls of much of the modern economy would lose their source of power, and, in some cases, substantial bribes. These changes would be required regardless of the form of ownership.

While technical production alternatives exist for much of the manu-

TABLE 2
INVESTMENT PER WORKER ASSOCIATED WITH ADVANCED
AND APPROPRIATE TECHNOLOGIES
(thousands of dollars per worker)

PRODUCT	ANNUAL OUTPUT OF PLANT	ADVANCED TECHNOLOGY ($000s)	APPROPRIATE TECHNOLOGY ($000s)
Shoes	300,000 pairs	2.2	0.8
Cotton weaving	40,000,000 square yards	37.6	8.7
Cotton spinning	2000 tons	14.7	2.0
Brickmaking	16,000,000 bricks	45.8	3.3
Maize milling	36,000 tons	9.7	2.9
Sugar processing	50,000 tons	6.2	0.8
Beer brewing	200,000 hectolitres	18.3	12.1
Leather processing	600,000 hides	36.2	15.5
Fertilizer	528,000 tons of urea	137.6	122.3

SOURCE See Table 1

facturing sector, the complementary program to obtain the benefits is quite radical, for socialist as well as market economies. Although aggregate gains could be realized, each of the groups just enumerated will perceive itself as losing relative to other social groups and will not accept such losses without an intense political battle. In contrast, the probable beneficiaries are too numerous and the benefits too uncertain to induce the formation of effective advocacy groups. The potential gains from the reform constitute a classic example of a collective good from whose benefit it is difficult to exclude people, and hence no individual perceives it to be in his interest to share in the costs necessary to realize the goal. Small wonder that few countries have systematically chosen the price realignment route, and those few often only after a combination of external pressure from suppliers of foreign aid and internal stagnation.

Skills and investment

I turn next to the question of whether the technical options that clearly do exist are quite as simple to implement as has been implicitly assumed previously. I will consider briefly two obstacles out of a half-dozen that may be important. These are, first, the possibility that less capital-intensive technologies, though employing more unskilled laborers, require a greater percentage of skilled labor and, second, that the information costs of learning about technological alternatives are substantial and thus a firm may find it more profitable to pursue other less expensive options to obtain increased profitability.

Among the more venerable platitudes invoked by LDC firm managers and bureaucrats, aid-giving agencies, and more than a few academic economists in defense of the choice of advanced technology, is that appropriate equipment requires more skilled operatives, maintenance workers, and/or supervisory abilities. Since all are in short supply, it is concluded that more modern equipment that economizes on all of these is desirable. The empirical basis for this view is rather tenuous, relying primarily on a few anecdotes that are often used as the medium of exchange at afternoon tea. Assume, nevertheless, that the view contains some substance.

Should this be raised to high principle and the corollary deduced that production with less advanced technology is impossible? All too often precisely this leap is made, ignoring the possibility that even if greater skills are required, they may be acquired by private or public expenditure on the relevant training. The cost of this investment must then be compared with the benefits to be obtained from the appropriate technology. While private firms may hesitate to make such a calculation, as workers they train may be pirated away by local firms or fly to Qatar, the refusal of LDC governments, public enterprises, and international agencies to pursue the benefit-cost calculus is shortsighted, as is the neglect of the usefulness of subsidies to encourage private firms to consider such training.

Using the data gathered in the studies that form the basis of the calculations presented in Tables 1 and 2, the benefit-cost ratios have been calculated for two of the industries in which the skill requirements associated with the appropriate technology are in fact greater than those required by the advanced technology. The ratio of benefits to costs obtained from investing in training is considerably above 30, whereas a ratio of 1 constitutes a justification for most projects. This result suggests that even where skill shortages are currently a factor limiting the adoption of labor-intensive technology, the desirable strategy for policymakers is to advocate a bundling of the requisite education and investment funds rather than passively accepting the adoption of unnecessarily advanced technology.

Cost of information

I now turn to the cost of acquiring technical information, a question that typically has not been emphasized in this context but is clearly of considerable importance in understanding a number of observed phenomena. Usable information about production alternatives, including machine and product specifications, raw material and power requirements, and typical complements of labor of various skills, is generally not readily available despite the textbook simplifications all economists use to represent technical choices. The studies on which Tables 1 and 2 are based, as well as many other studies demonstrating the existence of a considerable variety of technical alternatives, required the cooperation of a group of economists and engineers for a year or more. It is likely to be quite expensive to ascertain the relevant technical options among which a firm may choose.

It is easiest to obtain information about one or two technologies from presentations by manufacturers of capital goods. However, in many LDCs salesmen arrive only from the largest producers in the technologically most advanced countries, such as Switzerland and West Germany. Few representatives from Japan, Korea, or India visit the typically sparse markets of sub-Saharan Africa or the poorest countries in other regions despite the likelihood that some of their equipment may be suitable—hence the reference to the East in the first paragraph of this article. In general, it is time-consuming and expensive to determine relevant technical alternatives —attendance at trade fairs, careful examination of large numbers of trade publications, and ascertaining the performance in operation, as contrasted with the specification of machine manufacturers, from understandably secretive competitors in one's own country or abroad

—all require large monetary outlays or a considerable expenditure of managerial time.

While the hiring of consulting engineers might be thought desirable, they are unlikely to accept such assignments unless the total outlay on new plant and equipment is very large, as their compensation is a percentage of the amount of equipment purchases. Moreover, evidence is accumulating that they rarely have the breadth of knowledge to augment significantly the knowledge of even a mildly competent firm. Plant managers themselves will be reluctant to allocate much of their own or staff time to the necessary search unless the prospective payoffs—in terms of reduced costs—are very large, since many alternative uses exist for their time, including improvement of current levels of efficiency, finding lower-cost suppliers of raw materials, and eliciting more favorable treatment from all-important government agencies.

The role of information permits an explanation of the good performance of MNCs with respect to technology. They can more cheaply identify relevant machinery and transfer it among subsidiaries, particularly equipment which is losing its competitive edge in countries with high and growing wages but which would be appropriate in low-wage countries. The parent company may even have established a new plant partly to utilize such equipment in the production of exports from a low-wage country. Alternatively, the local MNC manager may request that the purchasing office of the parent company perform the search for desirable equipment. Such low-cost—to the local subsidiary—searches clearly increase the probability of the subsidiary's purchasing appropriate machinery that will allow it to take advantage of the relatively low price of labor in the LDC.

CONCLUSION

The obstacles, then, to the choice of appropriate technology and the realization of the large benefits it can confer are the absence of correct factor prices, a noncompetitive industrial sector sheltered by tariffs, information costs, and a tendency to perceive remediable shortages as impassable obstacles. This is nothing other than a short list of more general development problems that also afflict the larger and typically more important agricultural sector. Although the economic and technical policies to address many of these deficiencies have received careful analysis and are well understood, their implementation is not easy. There simply is no adequate understanding of the political economy of reform, whether it involves the reduction of subsidies to government enterprises or to consumers, or the reduction of tariff rates on excessively protected industrial sectors. Though there may be gains for the society as a whole, individuals or organizations—including here government bureaucrats and their agencies—incur losses relative to others, and there is no readily identifiable scheme that would allow the gainers to compensate the losers. The problems are common to all political systems: socialist Tanzania has shown, if anything, poorer performance than capitalist Kenya, and the development experience of China is not all that different from India's.

Many of the problems are symptoms of what Michael Lipton has called the urban bias in development, but, like a viral disease, this syndrome is easier to identify and name than to cure.[6] The continued problem of inappropriate technology in the industrial sector is one of several symptoms of the underlying malady. Unfortunately, changes in perception by policymakers and the acquisition of the knowledge and skills needed successfully to implement the required reforms are likely to develop considerably more slowly than recent progress portends for antiviral drugs. The art of political and economic management is not likely to witness the breakthroughs now perceived to be imminent in genetic engineering.

6. M. Lipton, *Why Poor People Stay Poor: Urban Bias in World Development* (Cambridge: Harvard University Press, 1977).

Technology Choice and the Distribution of Income

By GUSTAV RANIS

ABSTRACT: This article first explores the meaning that attaches to the flexibility in the choice of technology and defines the concept of equity in the distribution of income. It then proceeds to demonstrate how the ability to explore the limits of technological flexibility, both in its static and dynamic senses, can be linked to alternative distributional outcomes. The harshness of the typical import substitution sub-phase of development, the extent attention is paid to agricultural productivity growth, to the decentralization of industry, to the search for efficient labor intensity in both product mix and process choice, are all shown to make a substantial difference to the strength of the potential positive relationship between technology and equity. The real world experience of the East Asian developing countries that have managed to avoid conflict between growth and distribution largely via efficient technology choices is cited and contrasted with the more typical Latin American cases.

Gustav Ranis is professor of economics at Yale University, where he directed the Economic Growth Center for a number of years. His work has covered all aspects of development economics, with particular regard to Asia, and involved a number of years of overseas experience. He is the author of many articles and books in the field of development. Dr. Ranis received a B.A. from Brandeis, and a M.A. and a Ph.D. from Yale. He has been a Social Science Research Council and Ford Foundation fellow, and a frequent consultant to U.S. and international development agencies. He served as assistant administrator for policy and planning in the Agency for International Development (AID) from 1965 to 1967.

DURING the first two postwar decades growth of per capita income emerged as the main societal objective in the Third World. During that period overall performance was assessed relative to the two-and-a-half percent annual growth target, set by the UN and the donor community, a target which was exceeded, at least on the average—that is, taking all less-developed countries (LDCs) together. This creditable performance has, however, by no means been duplicated in the realm of the distribution of income, which remained poor, or even worsened, in most parts of the developing world.

The conventional wisdom of the 1950s and 1960s was, in fact, that countries must grow "first" and should worry about distribution "later." Implicit in this was that they must allow distributions to become worse, at least for a period of time. The cross-sectional evidence showing the worst distribution obtaining in middle-income countries seemed to underline the inevitability of it all. The prevailing argument was that as growth proceeds, rent and profit shares rise at the expense of wage shares kept low by the overhang of labor surplus conditions. In each sector the rich accumulate more. Moreover, there is a shift of families from the more egalitarian, agricultural to the less egalitarian, urban industrial sector. Historical evidence from most Latin American and some Asian countries—for example, the Philippines—seemed to confirm the general view of an inevitable conflict over some considerable stretch of time. The general notion was that if countries could only hold out until their rising saving rates propelled them into modern growth, the combination of higher wages induced by a new labor scarcity, plus fiscal redistribution, would eventually redeem the situation.

During the late sixties, and especially the seventies, this counsel of patience, however, began to be revised, partly because of an increased concern with the overall quality of life, absolutely and relatively, and partly because poor people's faith in the ultimate political willingness and technical capacity for redistribution began to waver. Moreover, there appeared on the scene a few real-world counterexamples or "deviant" cases in which equity actually improved with the acceleration of growth, all of which led people to question the inevitability of any "iron law." The empirical facts are that income distribution remained generally at unfavorable levels in most of the Third World during two decades of unprecedented high growth. The typical experience, in fact, was one of large increases in income per capita overall, associated with smaller increases in the incomes of the poor, making them relatively worse off than previously. Whether or not absolute poverty, in terms of the percentage of LDC populations below some arbitrary poverty level, in fact fell or increased is a more controversial subject. But it is generally agreed that the reassessment of the late 1960s and 1970s has addressed itself to at least including income distribution and poverty alleviation squarely among the objectives of contemporary LDC planners and policymakers.

This development task is shared by all societies in the Third World regardless of the particular social system they have chosen for themselves on the spectrum between really nonexistent pure market economies and equally nonexistent pure socialist economies. The ability of a society to achieve success in minimizing the conflict between growth and distribution objectives, moreover, clearly relates both to the initial conditions in which it finds itself and the policies it brings to bear over time. Looming large among the latter are those affecting the nature of technology choice and the direction of technology change. Together, the options utilized or discarded here are likely to make a tremendous difference with respect to the basic issue—that is, the extent to which growth and distribution can be made complementary and mutually reinforcing, rather than competitive, societal objectives.

In this article I hope to define and analyze the relations between technology and the distribution of income in the typical LDC setting. I begin with a few simple definitions and then sketch out what is approaching the current conventional wisdom relating to the choice of technology and its relevance to the avoidance of conflict between growth and distribution.

DISTRIBUTION AND TECHNOLOGY DEFINED

Income distribution refers to the ways in which the disposable income of a society accrues to the families in that society as a consequence of these families using their property and their manpower to produce income, plus transfers that may be arranged by government authorities. A perfectly equal distribution of income would occur if the disposable incomes for all families were spread exactly evenly, with all families receiving a proportional share of the total. And when the distribution is less than perfectly equal, as it is in every real-world case, some families receive less than their proportional share and some receive more. The degree of inequality can be conveniently illustrated by a diagram called a Lorenz curve summarized in a numerical form by means of the so-called Gini coefficient.

Figure 1 presents a Lorenz curve with a heavy black line running from A to B. The curve is constructed by plotting the percentage of the population or households along the horizontal axis, and the percentage of income received along the vertical axis. In this way the percentage of income received by each percentage of the population can be identified until 100 percent of the income and population has been accounted for. For example, in the example of Figure 1, the Lorenz curve indicates that the poorest 25 percent of the population received 10 percent of the total income, and that 75 percent of the population received 50 percent of total income. Perfect equality would produce a Lorenz curve along the diagonal broken line linking A and B, indicating that 25 percent of the population received 25 percent of total income, and so on. In fact, any Lorenz curve can be summarized as the ratio of the shaded area, which is the deviation from equality, to the area of triangle ABC. This number is called the Gini coefficient and clearly must lie between 0 in the case of perfect equality—that is, when the shaded area tends to vanish—and 1 for perfect inequality—that is, when the shaded area approaches the axes. Gini coefficients typically, in fact, range between .20 for the most equal cases and .55 for the least equal cases.

In analyzing the reasons for particular distributions of income in a particular developing country at a particular point in time, it is useful

Figure 1: Lorenz Curve

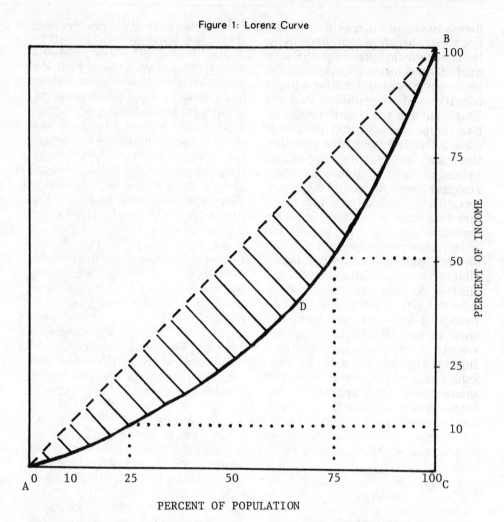

PERCENT OF POPULATION

to differentiate between the income distribution that results from the way income is generated via the production process and the disposable income results that are achieved thereafter as a consequence of redistribution either through private—charitable—or public—fiscal—transfers. In the mixed-economy developing country context, on which I focus here, it is the way output is generated that is likely to determine in very large part the final distributional outcome; and the way output is generated depends, in turn, largely on the technology choices made, both in terms of the selection of imported or transferred technology and in terms of the environment for appropriate domestic adaptations and indigenous innovations.

The reason for this relatively heavy reliance in developing countries on what may be called the "primary income distribution mechanism" is related to the fact that there is little tradition of private transfers between families and that the capacity of Third World governments to engage in public—that is, fiscal—transfers from upper- to lower-income groups after the production dust has settled is likely to be severely limited, both for administrative and sociopolitical reasons. This is particularly true of redistributive tax policies, poverty programs directed at the tail of the distribution, and public works programs intended to mop up large-scale residual unemployment, all activities difficult to organize, blueprint, and maintain in a mixed economy context.

India's experience with "post-office socialism," as John Kenneth Galbraith has called it, indicates that every time mixed-economy governments intervene directly on behalf of the poor, the poor find themselves worse off. What usually results is so-called program hijacking by the middle classes, some intentional, some the result of inadequate administrative, planning, and implementation capacity. While governments can admittedly help by reallocating expenditures, such as on health, education, and other public goods, from the urban rich to the rural poor, there is growing agreement among analysts that it is in large part the way output is generated over time and the way families are compensated for their contribution that will inevitably tell most of the story.

The two questions that are most important here are how well distributed the initial endowments are, including the distribution of land and the distribution of capital among the families; and second, given these initial conditions, what is the setting in which growth is permitted to take place, including the all-important issue of the role of technology and of technology transfer in that context.

We will have little to say here about the basic initial conditions, except to acknowledge that the initial population pressure on the land, the natural resources endowment, country size, and so on represent crucial parameters with respect to a system's total problems and opportunities, including the potential role for trade and technology transfer, versus domestic balanced growth and domestic innovations, in advancing any mix of objectives. Similarly, it should be obvious, *ceteris paribus*, that the more equal the initial distribution of the chips families have to play with—that is, land, capital, and so on—the more equal the distribution of income. The same is true with respect to the distribution of entrepreneurial capacity, educational heritage, and past infrastructural investments, as between urban and rural and rich and poor. Thus land reform, capital levies, inheritance taxes, equal access to educational opportunities, and the like are bound to be helpful. But even when these are not very equitably distributed—as is the case in most LDCs—the role of technology in improving the final income distribution versus growth result is substantial. The main purpose here is to demonstrate the role technology can play in determining income distribution outcomes via the way output is generated, given the initial constraint of varying initial endowment conditions that are part of the historical heritage of any society.

During the early postwar period the implicit assumption frequently made was that the choice of technology—that is, the range of choice by which to produce a given commodity efficiently across countries varying substantially in resource endowment—was quite narrow. We have learned since that this, in fact, is not the case and that a wide range of possible processes, characterized by different combinations of basic inputs, exists in nature for producing a particular specified final good.

This range of process choice has been found to have three main dimensions. The first is the demonstrated ability to run a particular core production process with widely differing numbers of workers per machine hour along different production lines, adding up to substantial flexibility in what constitutes an efficient core technology. Second, it has been found that even when a production process itself is fairly narrowly constrained in terms of the engineering possibilities, there normally exists a wide range of choice in the ancillary processes, such as the movement of intermediate goods between machines, in raw materials receiving, and in final goods packing, crating, storing, and so on. Third, there usually exist viable alternatives in the way the productive plant is organized in terms of the use of subcontracting or putting out systems, the extent to which purchasing and marketing may be handled differently—for example, via a trading company mechanism.

A second dimension of appropriate technology choice, more neglected until recently, is the possibility of choosing slightly different product specifications to fill the same final needs of consumers and/or producers. This point is made most easily by reminding ourselves that a shirt may have a number of attributes or characteristics—its strength, warmth, and washability. Such characteristics may represent accidental by-products of the chosen production line, but they may also be subject to conscious choice; for example, varia-

tions in the characteristics of a given item can often occasion substantial additional flexibilities in the production process—a bush shirt may not only be more appropriate to a developing country's climate than a drip-dry shirt, but it may also be able to absorb many more unskilled labor hours per unit of capital. Thus the question of appropriate technology choices must be extended from the technical process to the combination of process and attribute mixes that are feasible.

We might, in other words, think of the final technology in place in a given developing country as a composite of large numbers of process and product choices, combining what is initially chosen, often via technology transfer, with indigenous innovation and/or domestic adaptation "on top of" the imported choice. The pure transplantation of technology from abroad, in either sense of the word, is likely to lead to lesser total responsiveness to the potential flexibility that exists in nature and to a lesser ability of each country to utilize its own particular resource endowment as fully as possible at each point in time.

We should emphasize at this point that by a range of potentially available technologies we mean efficient technologies relevant to a range of varying endowments and taste situations. We are excluding the choice of employment-generating processes and/or attributes in the "make-work" sense—that is, one which connotes a necessary sacrifice of output. Rather, we are considering the possibility of ending up with an appropriate technology in a cost-minimizing—or at least cost-reducing—sense, which, for the typical developing country, is likely to mean exploring the range of labor-intensive choices statically

Figure 2: Technology Transfer and Adaptation

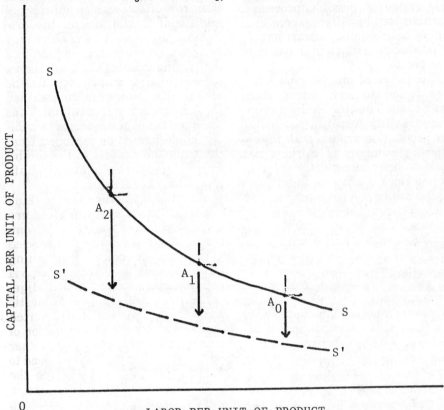

and moving in the direction of labor-using changes in technology dynamically.

The scope for such a range of alternative technologies in nature may be depicted for a hypothetical case in Figure 2. With only two factors of production—capital and unskilled labor—assumed to collaborate in producing output, the vertical axis measures the amount of capital needed to produce one unit of a given product and the horizontal axis measures the amount of unskilled labor needed to produce that same unit. The various technologies available, including internationally via technology transfer or borrowing, differ in the relative pro-portions of capital and labor used to produce one unit and occupy different positions or points on the graph as indicated. These points may be interpreted as pertaining to different countries of origin currently and/or combinations experienced in the same country at some earlier point in time. For example, technology A_0 may be the technology used in Germany in 1920 to produce a given commodity unit; A_1, the technology used in the United States in 1920; A_2, U.S. technology vintage 1950. Points on the lower righthand portion of the SS curve thus represent the more labor-intensive technologies, either in terms of earlier points of observation in the now advanced countries or current points of observation in the intermediate-level countries. The full range of choices can, of course, be imagined to widen further once the specific attributes or characteristics of the given unit of output are permitted to vary.

Much of the technology in use in the developing countries—certainly in their modern sectors—originates in the developed countries. This is less true in agriculture, but even there much new knowledge is generated in the context of international research institutes. But virtually everywhere the final choice of technology is customarily one of some blend or combination of technology transfer from abroad and domestic adaptation. The transfer, in terms of our SS curve in Figure 2, relates to the potential availability from the international-cum-historical shelf of knowledge, that which is most suitable to the country's own endowment. But in addition to this potential choice of technology from the so-called available international shelf, and perhaps more important, is what may be called innovational assimilation—that is, innovation "on top of" the domestically known or imported technology, presumably in the direction of using relatively more of the abundant unskilled labor supply than in the country or international institute of origin. It is important to realize that no technology developed for use under one set of conditions can ever be used under another set of conditions without some degree of adaptation.

The growth of the scientific and technological capability to make appropriate changes of this type is, however, crucial. Such indigenous "capital stretching" or the devising of innovations in a labor-using direction can be represented as a reduction in the capital coefficient per unit of output for each of the technology choices referred to in Figure 2. The effective postassimilation set of unit technologies—that is, after domestic assimilation—can thus be illustrated by some curve like S'S', with the strength of the labor-using innovative effort indicated by the amount of downward shift in the capital coefficient permitting more workers to be employed per unit of capital.

One task that remains is to provide a conceptual link between the technology choices finally made and the income distribution that results. The other is to sketch in some of the reasons why, at least in a minority of actual contemporary LDCs, this link was important in setting aside the customary pessimism with respect to the relationship between growth and the distribution of income.

LINKAGES BETWEEN DISTRIBUTION AND TECHNOLOGY

The extent to which the changing endowments of a society—that is, the various factors of production "owned" by families—are employed in the course of growth may be seen as the most important single determinant of income distribution results in the course of the transition growth effort. This, in turn, is in large part a function of a society's ability to adapt its technology to changing conditions as it moves through various phases of growth— for example, from infant-industry-dominated import substitution, usually characterized by heavy government intervention, to export substitution, usually characterized by much greater market orientation. The adoption of appropriate processes or products by individual entrepreneurs is closely related to the overall policies followed by the society that either permit a society's resources endowment to be "heard" via relative price signals at the individual entrepreneurial level or serve to mute the impact of such forces on the decisions made in a large number of dispersed units in both the agricultural and industrial sectors.

These choices are well demonstrated by the sharp contrast between East Asian countries such as Taiwan and Korea, which have moved resolutely toward more export-oriented, endowment-sensitive sets of technology choices, and Latin American countries such as Mexico or Colombia, which have basically persevered in import substitution policies. While I will only indicate the general connections here, such divergencies in policy and growth patterns can be causally linked to differential equity performance. Different income distribution outcomes result directly from the way technology and product mix decisions are made, given a society's initial conditions.

While there is no need here to go into the technical details of that linkage, a useful device to attempt to relate varying distributional outcomes to the nature of the growth path is by decomposing our Gini coefficient, discussed previously, into its various factor income components. In an economy that has a substantial agricultural sector it is, moreover, essential to differentiate between rural and urban households because of differences in the organizational and spatial dimensions of economic activity that physically affect the way the primary output decisions are made and income distribution results occur. While urban families are engaged mainly in industrial and service activities, rural families are engaged in agricultural activities as well as in nonagricultural activities that generate wage and property income. In other words, rural family income from agriculture can be separated from rural family income from nonagriculture, and further divided into wage and property income, with each component weighted by the equity of its own distribution and its own relative importance in the families' total income.

Thus we can think of the income inequality of a society at one point in time, as well as over time, as the weighted sum of changes in the distribution of various components and in their respective weights. And these, in turn, can be directly related to the way in which output is generated and process and product choices are made in the course of the development process. For example, an increase in the share of labor income, which is usually more equally distributed, and a decrease in that of property income, which tends to be relatively less equally distributed, is usually favorable to overall income distribution as measured by the Gini coefficient. This, in turn, is a function of a more labor-intensive or employment-intensive set of process and product choices.

Taiwan, in fact, provides one of the most striking and best-documented cases in which a remarkably strong growth performance was combined with low levels and continuing further decreases in income inequality. The Gini for rural families' agricultural income, for example, continued to decline not only due to an initially favorable distribution of assets, post-land reform, but also because technologies were developed and promoted that made small farm holdings more productive than large holdings over time. Land was used more intensively by choosing such new labor-intensive crops as mushrooms and asparagus to augment rice, by replacing the more land-intensive sugar, and by cropping the same land several times. Such technological changes were particularly beneficial to the poorer farmers, who were able to participate more than proportionately.

Also, with respect to the same rural families, the decentralized nature of the industrial structure, with a large number of small-scale rural enterprises providing up to 50 percent of rural family income over time, was an important component of the link between technology choice and an increasingly equitable distribution of income. When control over scarce inputs such as foreign exchange, credit, and so on diminishes, as it does when a country moves from import substitution toward export orientation, it is much easier to avoid discrimination against rural, medium- and small-scale enterprises that, in combination with the existence of rural infrastructure, permit poorer farm families to augment their income from agriculture with increasingly important income from rural nonagricultural service and industrial activities.

Such a decentralized industrialization pattern, as witnessed in Taiwan, has the following advantages. Industries provide more by-employment and higher incomes, especially for the poorest rural families; transport costs for the economy as a whole are lower than those incurred by an alternative, concentrated pattern; and the costs of developing infrastructure for a dispersed small-scale industrial sector are much less than for the large, concentrated urban centers. Any ideal or equilibrium market area always reflects a compromise between economies of transport costs favoring smaller market areas and economies of scale favoring somewhat larger market areas. Thus an output bundle initially focusing on industries like textiles and other labor-intensive nondurable consumer goods that do not have pronounced economies of scale is favorable to the complementary interaction between technology, growth, and the improvement of the distribution of income.

In addition, the proximity of a modern nonagricultural sector in the rural areas contributes to the modernization of agriculture both in the incentives and the input-output or technological linkage sense.

There exists clear evidence that the poorest families more often seek and profit from the by-employment opportunities offered by such decentralized rural industry and services. It is a combination of appropriate goods being produced and traded in the rural areas and appropriate—that is, labor-intensive—technologies being adopted that can lead to an extraordinary absorption of disguisedly unemployed labor hours in an efficient fashion. This factor made a tremendous contribution to the improvement in the distribution of income in such fast-growing countries as Taiwan during the 1950s and 1960s.

A third important potential contribution to the lowering of the overall Gini coefficient is via the level and trend of the labor income and property income shares relating to the incomes of urban families. A typical Latin American case, for example, has the share of urban family labor income in total income between .2 and .3 and declining, during the 1960s, while that of the typical East Asian country is around .4 and rising, reaching almost .6 by the early 1970s.

All this represents evidence of the more labor-intensive process and product mixes in the East Asian cases. A growth strategy that entails greater sensitivity to factor endowments and that—once the initial period of infant industry protection has run its course—encourages penetration of foreign markets with labor-intensive nondurable consumer goods, under less distorted policy regimes, is much more able to absorb workers productively and thus to permit a stronger link between appropriate technology choice and an improving distribution of income. There can be little doubt that the increase in labor's relative share, both urban and rural, contributed powerfully to low and declining overall Ginis in the East Asian country cases, while the continuation of import substitution policies in much of the rest of the developing world was a major contributor to the relatively much lower resort to labor-intensive choices and to the continued higher levels of income inequality. This contrast illustrates the importance of how output is generated and what growth path is selected—that is, the extent to which a society's endowment can be accommodated as a consequence of a flexible choice of technology in the broadest sense of that term—rather than how much growth can be achieved.

The distortion of relative factor prices and of relative commodity prices, which often goes hand in hand with the so-called import substitution policy syndrome, is crucial here. Such price distortions affect not only the product mix but also the choice of technology in terms of process and product quality because they tend to favor capital-intensive, urban-oriented, and large-scale industries. While such policies have the objective of providing infant industry protection and are intended initially to shift foreign and domestic resources into the new, rapidly growing consumer nondurable goods industries that supply domestic markets with previously imported materials, they often result in import imitation rather than replacement and in the use of unadapted or turnkey imported technology. When this phase of development comes to its inevitable end, LDCs generally begin to produce previously imported capital and durable consumer goods and to process their own raw materials, all implying a continuation, and often even an intensification, of these same policies.

What is perhaps less well understood is the importance of phenomena that usually accompany this distortion of relative prices—for example, the impact of windfall profits on industrial sector growth, on competitiveness, and on the extent of pressure to choose the most appropriate technology in both its process and attribute mix dimensions. Import licensing systems, overvalued exchange rates, official low interest rates for favorite borrowers, the preferential allocation of scarce raw materials, and the like, not only distort relative commodity and factor prices away from appropriate technologies, but they also create an environment that favors satisficing rather than maximizing behavior by the large-scale urban industrial firms.

Once the economy emerges from its initial, natural-resources-fuelled, import-substitution growth phase, it must decide whether to move into an unskilled-labor-based, export-oriented subphase—the East Asian case—or to continue with import substitution, but now shifting toward durable consumer and capital goods production—the Latin American case. In one case, increased sensitivity to the changing factor endowment comes into play both in terms of the efficient utilization of resources statically and in terms of the incorporation of growing domestic ingenuity in what is being produced and how. In the other, the veil between endowments and prices thickens, if anything, while entrepreneurial energies are even more focused on obtaining a favored place in the queue for directly allocated "goodies," rather than scurrying about in the search for the construction of "better mouse traps."

If a government has somehow decided to shift to a relatively catalytic rather than a directly interventionist or dirigiste role, this materially improves the chances that millions of dispersed decision makers in both sectors of the economy will be motivated to search for more appropriate technological choices. In economists' jargon, the narrowing of the gap between market and shadow prices means that transfer or transplantation choices become more flexible or labor-intensive and, more importantly, that the assimilation or adaptive type of innovational behavior is likely to be more pronounced and in the "right" direction. The search for indigenous innovation becomes a conscious activity of the individual entrepreneur only if the environment is conducive—that is, if more realistic relative price signals and enhanced competitive pressures are

likely to combine to take the form of indigenous "capital stretching" on top of the imported technology.

Why, it might well be asked, do not other developing countries follow the contemporary East Asian example—shared, incidentally, by Japan at an earlier stage: achieve faster growth via a more endowment-sensitive set of technology choices? A Toynbeesque answer might well be that individuals, like societies, seem to have a tendency to follow the road of least resistance. Accordingly, the more typical developing country—for example, in Latin America—has traditionally been biased toward a natural-resources-based growth and export pattern and has continued to rely heavily on such exports in the course of its transition growth effort.

This pattern has translated itself into both a relative neglect of food-producing agriculture and the avoidance of some of the painful macro-policy changes that might lead to a fuller utilization of unskilled human resources by the end of import substitution. A similar capacity to prolong the import substitution industrialization phase via the exploitation of some sort of natural resources bonanza did not exist in Japan or in the contemporary East Asian countries. Thus these countries were unable to skip the labor intensive export substitution phase, which is so heavily linked with the appropriateness of technology choice and the direction of technology change in a labor-using direction.

It does seem as if the availability of more natural resources and/or more capital inflows, public or private—both of which should be helpful in easing the policy transition pains from one policy regime to another—can just as easily be used to avoid unpleasant changes along the growth path. While "nature refuses to make jumps," the basic issue of ensuring complementarity among societal goals may have much to do with the ability of governments who are not "up against it" to persuade themselves and various interest groups that appropriate changes in policy may nevertheless be in everyone's interest. Industrialists might be better off—may make larger profits—in a more competitive and endowment-sensitive production and export context; working families might be better off focusing their attention on larger total wage incomes resulting from enhanced employment opportunities, rather than higher wage rates for its few members employable as part of the organized labor force. And while we are trespassing here on the realm of political economy, the formation of coalitions among such groups to effect policy changes is probably necessary if the full potential of technology is to be enlisted on behalf of a more equitable transition growth path in the developing world.

Benefits and Obstacles to Appropriate Agricultural Technology

By ROBERT E. EVENSON

ABSTRACT: Most forms of agricultural technology have not been transferred from developed to developing countries. This is particularly true of technology of the biogenetic type. Plants and animals interact with the soil and climate environments in which they live. Natural selection pressures produce highly varieagted species and types of plants and animals, each suited or appropriate to an environmental niche. Modern plant breeding methods have only partly overcome the sensitivity of biological material to environments. Environmental interactions play a lesser role in biochemical and mechanical technology in agriculture. Research programs to improve agricultural technology accordingly require a high degree of targeting to local environmental conditions. In some cases new technology, as in the improved "green revolution" wheat and rice varieties, is initially adapted to a wide range of environments. Improvements on the initial high-yielding varieties have tended to be more narrowly adapted. Agricultural research programs serving relatively small regions have been successful in undertaking adaptive research to develop region-specific improvements to varieties developed in international centers in the developing countries.

Robert E. Evenson is professor of economics at Yale University since 1977. He has been a member of the faculty of the University of Minnesota and a visiting professor at several universities in developing countries—particularly at the University of the Philippines in Los Banos. He has published a number of papers investigating the transferability of agricultural technology.

AGRICULTURAL production is based on biological processes. Both plant and animal commodities require the growth and reproduction of living organisms. These organisms are subject to disease and insect problems. Their growth processes are affected by differences in soil qualities, temperatures, water availability, and a number of other environmental factors. Appropriate agricultural technology thus encompasses a biological dimension as well as the mechanical and chemical dimensions dominating many other classes of production technology.

In this discussion of the benefits and obstacles to appropriate agricultural technology, I will be specifically concerned with the developing regions of the world and with the possibilities for transfer of technology produced in other regions to these regions. The relevance of the biological process dimension to the issue is so dominant that it requires attention first. After considering both the premodern and modern approaches to plant and animal improvement through selective breeding, I will discuss the chemical and mechanical dimensions. Finally, I will review the evidence regarding the effectiveness of research by agricultural scientists to produce more appropriate agricultural technology for different parts of the world.

BIOLOGICAL PROCESS COMPONENT OF AGRICULTURAL TECHNOLOGY

Charles Darwin, in his famous *Origin of the Species*, established a number of principles guiding the relationships between types of biological technology and the environments in which they have a comparative advantage. The fact that the world's surface provides a rich variety of environments—environmental niches—in which plants and animals grow has been well documented for centuries. Similarly it has long been known that an incredibly rich array of differentiated species of plants, animals, insects, parasites, and pathogens has been associated with these variegated environments. Darwin clarified this relationship between species and environments with an argument that has economic characteristics. The survival of species is a matter of appropriateness of biological technology to environmental conditions. Given that natural mutation processes were continually producing new genetic variation in living matter—that is, new technology—there was a natural sorting out of the "fittest" for each environmental niche.

Human populations have produced an economic system of plant and animal husbandry designed to alter the natural system in order to produce more valuable products. Important and economically valuable plants and animals are domesticated.[1]

This altered the Darwinian equilibrium in several ways. First, the valued species evolved through time under husbandry selection pressures. Second, in response to this selection, the equilibrium was altered regarding nondomesticated species. Pests, parasites, and pathogens associated with cultivated crops often found improved environ-

1. It is of some interest to note that, with only one exception, all modern crops and animals were domesticated from preexisting material centuries ago. The one exception is triticale, a crop originating in the early twentieth century from the work of modern plant breeders.

ments in which to survive as new types of crops were developed.

The development of crops and animals over the period of husbandry selection, which lasted until the nineteenth century, when modern scientific breeding methods were utilized, shows a pattern of changing comparative advantage of regions for crop production. Husbandry-selected technology tended to be adapted to environments other than those best suited to its origination. Virtually all modern crop and animal species and types eventually had better economic performance outside their "centers of origin" than in the regions where they first emerged. This appears to be the result of selection and reproduction which allowed crops to "escape" from their most serious pests and pathogens. Since most centers of origin were in tropical and subtropical climate regions, the temperate zone regions of the world were favored during the premodern period of plant improvement.

The modern period in agricultural science dates from the early 1800s when the field of agricultural chemistry was established in Germany and the Rothamstead Experiment Station in England began the systematic application of scientific methods to agricultural production technology. This early work moved agricultural improvement efforts out of the "country gentlemen" societies and the botanical gardens and into the laboratories. By the early 1900s hundreds of agricultural experiment stations had been established in many parts of the world. The United States had by then a well-established State Agricultural Experiment Station system. The European countries and Japan also had well-developed research pro-

grams. In today's developing world, with few exceptions, the only significant research programs prior to 1950 were directed toward the improvement of the "colonial crops": sugar, tea, coffee, cocoa, and cotton.[2]

In general the developed countries of the world entered the modern period with a large comparative advantage in most crop and animal production. Except for a few crops specific to the tropics, yields per acre were higher in temperate zone countries. This advantage was greatly increased with the advent of modern agricultural science, for two reasons. First, the temperate zone developed countries invested in agricultural science and built effective research systems, while the developing countries did not— except for the colonial crops. Second, modern agricultural science only partly overcame the fundamental linkage of technology to the environment that characterized premodern development. As a consequence, very little of the technology produced in the developed countries was actually transferred to the developing world.

The strength of the technology-environmental linkages has been persistently underestimated by policymakers. In the early years of the modern period of emphasis on economic development, policymakers emphasized agricultural extension and community development programs as the means to rapid development. Agricultural extension advisors swarmed over the developing world in the 1950s, bringing U.S. and European "know-how" to the farmers of the tropics. By the

2. See James K. Boyce and Robert E. Evenson, *Agricultural Research and Extension Programs* (New York: Agricultural Development Council, Inc., 1975), pp. 78-100.

early 1960s, it was clear that U.S. know-how, including virtually all aspects of agricultural technology— varieties, machines, and even chemicals—was simply not transferable to environments that differed greatly from those for which it was developed.

This led to a modification of the earlier development strategy. International agencies have, for the most part, pursued a bifurcated strategy over the past two decades. The emphasis on technology transfer has continued in the form of "rural development" projects which continue to receive the bulk of development aid. These projects are greatly varied in nature but often have a know-how transfer component. The second component of the strategy has been the support of agricultural research systems in the developing countries.

This research system development strategy has taken primary form in the building of the system of International Research Centers. The International Center for Wheat and Maize Improvement (CIMMYT) in Mexico and the International Rice Research Institute (IRRI) in the Philippines are the oldest and best known of these centers.[3] In recent years a number of national research programs have also attained significant research capacity, and current international policy

is slowly shifting toward further strengthening of these programs.

Tables 1 and 2 provide a summary of comparative research expenditures. Table 1 reports research spending as a percent of the value of agricultural product for five categories of countries, grouped according to per capita income in 1971. The disparity in spending patterns between rich and poor nations is readily apparent. It is also clear that this disparity is less severe for national public expenditures than for the total of national public, international, and industrial expenditures. Table 2 shows the commodity orientation of research spending in all developing countries; the data on proportion of product value are not fully comparable with the data in Table 1 because general research that cannot be associated with commodities is included in Table 1. The research emphasis of the international centers is shown in the table. It is clear that while some major commodities have reasonable research programs in place, others do not. The negligible research on important commodities such as cassava, coconuts, sweet potatoes, groundnuts, and chick peas is especially noteworthy. Cotton appears to be the only commodity in the developing world with research emphasis comparable to the emphasis placed on it by developed countries.

It is, of course, natural to ask whether, in the absence of an elaborate system of agricultural research centers, each serving a particular environmental region, the limited research capacity in the developing world might produce more widely adaptable technology. In other words, could a research center produce technology that would be worth more in environments other

3. International Center for Tropical Agriculture (CIAT), Palmira, Columbia; International Institute of Tropical Agriculture (IITA), Ibadan, Nigeria; International Potato Center (CIP), Lima, Peru; International Crops Research Institute for the Semi-Arid Tropics (ICRISAT), Hyderabad, India; International Center for Agricultural Research in the Dry Areas (ICARDA), Beirut, Lebanon; International Laboratory for Research on Animal Diseases (ILRAD), Nairobi, Kenya; and International Livestock Center for Africa (ILCA), Addis Ababa, Ethiopia.

TABLE 1

**RELATION OF WORLD EXPENDITURES ON AGRICULTURAL RESEARCH
TO THE VALUE OF AGRICULTURAL PRODUCT, BY INCOME GROUP**

GROUP	PER CAPITA INCOME (U.S. dollars)	PERCENTAGE OF TOTAL AND PUBLIC RESEARCH EXPENDITURES TO VALUE OF AGRICULTURAL PRODUCT*			
		1971 TOTAL	(PUBLIC)†	1974 TOTAL	(PUBLIC)†
I	1750	2.48	(1.44)	2.55	(1.48)
II	1001-1750	2.34	(1.76)	2.34	(1.83)
III	401-1000	1.13	(0.86)	1.16	(0.92)
IV	150-400	0.84	(0.71)	1.01	(0.84)
V	150	0.70	(0.65)	0.67	(0.62)

*Total expenditures include: (1) national public agriculture, (2) national public agriculture related, (3) industry, and (4) international. Excludes People's Republic of China.
†National public agriculture (national public agriculture related is not included).
SOURCE: Adapted from Table 1.7 in James K. Boyce and Robert E. Evenson, *Agricultural Research and Extension Programs* (New York: Agricultural Development Council, Inc., 1975), p. 11.

TABLE 2

**ESTIMATES OF INTERNATIONAL AND NATIONAL RESEARCH INVESTMENT
BY MAJOR COMMODITIES, 1971 CONSTANT DOLLARS**

COMMODITY IN ORDER OF VALUE OF PRODUCTION	VALUE OF COMMODITY IN ALL DEVELOPING NATIONS ($ billions)	ESTIMATED RESEARCH INVESTMENT		NATIONAL INVESTMENT AS PROPORTION OF PRODUCT VALUE (percentage)
		INTERNATIONAL CENTERS (1976)* ($ millions)	NATIONAL CENTERS (1976)† ($ millions)	
1. Rice	Over 13	7.9	34.7	0.26§
2. Wheat	5 to 6	3.8	35.9	0.65
3. Sugarcane	5 to 6	none	30.2	0.50
4. Cassava	5 to 6	1.9	4.0	0.07
5. Cattle	5 to 6	7.9	54.8	0.88
6. Maize	3 to 4	4.1	29.6	0.75
7. Coconuts	3 to 4	none	2.0	0.06
8. Sweet potatoes	3 to 4	0.6‡	3.4	0.09
9. Coffee	2	none	8.5	0.40
10. Grapes	2	none	6.9	0.35
11. Sorghum	1 to 1½	1.2	12.2	0.77
12. Barley	1 to 1½	0.5	9.4	0.62
13. Groundnuts	1 to 1½	0.5	4.0	0.13
14. Cotton	1 to 1½	none	60.1	3.50
15. Dry beans	1 to 1½	1.5	4.0	0.25
16. Chick peas	1 to 1½	1.2	3.0	0.18
17. Chillies and spices	1 to 1½	none	4.0	0.25
18. Olives	1 to 1½	none	5.0	0.33
19. Grain legumes	1	1.6	(25.3)	(2.00)
20. Potatoes (white)	1	2.0‡	8.2	0.68

*Centers and programs sponsored by the Consultative Group on International Agricultural Research.
†Rough estimate derived by allocating total research expenditures by country according to the proportion of standardized publications. Standardized publications are converted into constant scientist-years.
‡Additional funds also were spent on these crops at the Asian Vegetable and Research Development Center.
§The proportion varied sharply by type of rice: shallow water, .40; upland rainfed, .16; intermediate, .16; and deep water, .05. The international center investment was principally in the first two types.
SOURCE: Robert E. Evenson, Supporting Papers, vol. 5 (Washington, DC: World Food and Nutrition Study, The National Research Council), p. 51.

than those of its immediate location? The answer to this question is that in plant breeding there is scope for a tradeoff between adaptability and local effectiveness. The degree to which this tradeoff can be made varies according to the crop. In some crops, such as maize, there is little scope for wide adaptability; in others, the scope is considerable.[4]

The basis for this tradeoff exists because plants do differ in their tolerance of environmental factors such as temperature, humidity, and soil salinity. Breeders can select varieties that are more tolerant or, to use a modern term, have lower action. Selecting for low interactions usually means giving up some other desirable trait. Most advanced country research systems have had little incentive to seek aggressively widely adaptable materials because they have elaborate regional research systems designed to produce varieties for relatively small regions.

There are at least three important cases, each associated with a "green revolution," in which wide adaptability of plant material played a major role. In each case the supporters of the research system had motives that gave high weight to wide adaptability. Also, each case demonstrated another principle or facet of wide adaptability: when superior technology is made available to a region by transfer from another region, the receiving region has good potential to add to the value of the transferred material through local adaptive research.

The first of these cases is the development of improved sugarcane varieties in the 1920s. After the discovery of methods to induce the cane plant to flower and reproduce sexually in 1887—prior to this, all cane was propagated asexually, except for rare natural sexual reproduction —a number of improved varieties were developed in the experiment stations of the major producing countries. These modern varieties were susceptible to many local diseases and were not widely adapted. In the early 1920s, the experiment stations in Java (P.O.J.) and India (Coimbatore) developed "interspecific hybrids," which incorporated genetic material from hardy, noncommercial canes. This was almost an accidental production of widely adapted varieties, although the colonial interests of the period that supported this research were generally interested in improving technology in a number of countries.[5]

The early interspecific hybrids, especially P.O.J. 2878, were planted in a number of countries and quickly replaced local canes in much the same fashion as the Mexican semidwarf wheats and the semidwarf rices did some decades later. At one point in the 1930s, the P.O.J. 2878 variety probably accounted for 40 percent of the world's sugarcane production.

The widely adapted international varieties were initially superior to local varieties over a wide range of environments, probably in as much as 90 percent of the world's sugarcane area. Virtually every country or region which developed a local program to improve further on these varieties, and to "target" them

4. See R. E. Evenson, J. C. O'Toole, R. W. Herdt, W. R. Coffman, and H. E. Kauffman, "Risk and Uncertainty of Factors in Crop Improvement Research," in *Risk, Uncertainty and Agricultural Development* (Philippines: SEARCA, College Laguna, 1979).

5. For a discussion of this see R. E. Evenson and Yoav Kislev, *Agricultural Research and Productivity* (New Haven, CT: Yale University Press, 1975), pp. 34-57.

to local regions, however, was successful in doing so. By the 1960s, the world's sugarcane acreage was almost entirely planted with varieties which utilized the 1920s material in breeding programs but which were highly targeted to local environments. No single variety accounted for a large part of the world's production.

The development of the semidwarf wheats in the Mexican Rockefeller Foundation program—later to become CIMMYT—has clear parallels. The funders of this research gave high priority to producing new technology that would be widely available to countries that generally had not improved their own varieties and did not have a strong capacity to do so. Wide adaptability was emphasized, and screening procedures were devised to obtain it. The introduction of the first Mexican wheats in Pakistan and India in 1965 or so was followed by their rapid adoption. Within a short period those countries with a strong research capacity produced local adaptations to these international varieties. Varieties produced at CIMMYT probably accounted for 25 percent of the wheat acreage in the developing regions by the mid-1970s, but were being rapidly replaced by local adaptations. Dalrymple's most recent data suggest that the combined total of CIMMYT varieties and CIMMYT-induced adaptations accounts for more than 70 percent of the acreage for Asian countries.[6]

The third case, that of semidwarf rices, is similar. By the early 1960s the semidwarf technology was ripe

for development. In the Philippines the University of the Philippines College of Agriculture had already produced several important high-yielding varieties, particularly C4-64, which was bred at the same time as IR-8. The establishment of IRRI provided major new impetus to the development and spread of the high-yielding rice varieties. As with wheat, national research programs quickly incorporated the IRRI and other materials into local breeding programs and have now produced hundreds of locally bred semidwarf varieties. Since the semidwarf rices are suited only to environments with a high degree of water control, their maximum transfer potential across environments, even when locally modified, is to only approximately 50 percent of the rice-producing regions of the tropics and subtropics.

These three are not the only successful cases of development of appropriate technology in the developing world. Many more cases, including localized success in corn, have been documented. Corn, however, is subject to strong genotype-environment interactions, and this has blocked the development of anything parallel to wheat, even though CIMMYT has been pursuing a corn research program for many years. The cases discussed previously illustrate the natural evolution of at least some of the international centers. CIMMYT and IRRI have evolved into "wholesalers" of specialized technology and technology components.

CHEMICAL TECHNOLOGY IN AGRICULTURE

The major types of chemical technology of relevance to agriculture

6. See Dana Dalrymple, "Development and Spread of High Yielding Varieties of Wheat and Rice," Foreign Agricultural Economic Report No. 94 (Washington, DC: U.S. Dept. of Agriculture, 1978), p. 125.

are fertilizer, herbicides, insecticides, and animal pharmaceuticals.[7] Fertilizing materials have long been used in agriculture. Most inorganic fertilizers have been in production for a century or more. Major advances in efficiency of production and in handling technology have taken place, however. The real prices of most fertilizers have fallen over most of the twentieth century. During the 1950s and 1960s, price declines were quite dramatic for nitrogenous materials. Since most nitrogen-producing processes rely on oil or gas raw materials, prices have risen in the 1970s.

With declining fertilizer prices, consumption increased and agronomists and soil scientists developed more effective systems for applying fertilizers and minerals—chiefly lime and trace minerals—to compensate for soil deficiencies. The implementation of these systems was assisted by extension and education programs. Plant breeders also responded to declining fertilizer prices by placing more weight on fertilizer responsiveness in breeding programs. Both the high-yielding green revolution rices and wheats were selected and designed for high fertilizer responsiveness. Most of the yield increases in crops throughout the world in the past century have been associated with fertilizer. Fertilizer itself is not new technology. The complementary varietal improvements and managerial improvements are the sources of the yield improvements.

Insecticides, herbicides, and related chemicals have been important in the past three decades or so. Most of these chemicals have been

7. The biological technology referred to earlier is also biochemical in nature. The term "chemical technology" as used here refers to industrial chemical products.

developed by private industry rather than the public sector, which has dominated biological technology development. Herbicides have, for the most part, been economically important only where labor costs are high. In general, hand labor can achieve the same or better control of noxious weeds in most situations; thus we find that they have little impact on most of the tropical developing countries.

Insecticides, on the other hand, can achieve pest control not possible by hand labor. They are important in the developing countries, particularly when new crop varieties are introduced. The experience with many improved varieties is that they are often susceptible to pests and disease attacks—often unimportant and unnoticed pests and disease—within a year or two after introduction. The brown plant hopper, which became a serious problem in rice production after the introduction of IRS, had not previously been a significant pest because older rice varieties had cell characteristics that gave it resistance. Chemical methods of control were utilized but in this case were not highly effective.

In a remarkably short time breeders at IRRI were able to develop a cross between IR-8 type material and genetic material from its genetic collection resistant to brown plant hopper attacks. Many thousands of varieties were screened before a resistant noncommercial source was found. Within a year or so after first indications of the severity of the problem, IRRI released new resistant varieties. Today IRRI screens its varietal materials for resistance to several diseases and several insects, including four "bio-types" of the brown plant hopper. As a consequence, the importance of chemical control has been lessened.

In the animal health field, however, genetic improvement has not been as effective a substitute for chemical or pharmaceutical control. The sensitivity of animals to environments is not always fully appreciated by students of technology. One has only to travel short distances in the tropics to note marked differences in the size and strength of work animals living in uncontrolled environments. The work horse, for example, does not thrive in the tropics. Work cattle vary greatly in size, and in some regions native cattle are too small to be useful work animals. Yet they are the only types that can survive in particular insect and disease environments.

MECHANICAL TECHNOLOGY IN AGRICULTURE

Machines for land preparation, weed control, and harvesting and processing of agricultural crops have long been the objective of inventive activity by farmers, blacksmiths, industrial firms, and public sector agencies. Prior to the nineteenth century, considerable development of plows, animal harnesses, and the like had taken place. The nineteenth century, however, was the age of invention for agricultural implements. The first patent granted by the U.S. Patent Office was for an improved plow in 1796. By the end of the century some 60,000 to 70,000 patents had been granted for hundreds of types of plows, cultivators, specialized planting machines, and weed control devices, and for numerous types of harvesting and threshing machines. Not only were thousands of patents granted for inventions of

new agricultural machines, but many were produced and sold. Paul David, in his analysis of the adoption of the reaper, notes that farm machinery was the largest industry in the United States by the end of the nineteenth century.[8] In the twentieth century a new series of inventions was induced by the development of the tractor as a power source.

It is relevant to ask why, throughout much of today's developing world, land is still prepared with a bullock pair pulling a simple wooden plow, grain is harvested with a simple scythe or hand-held knife, threshing is done by hand, and rice is hand pounded? Machines for all of these tasks have been under a continuous state of improvement for more than a century in other parts of the world. Why are they not being used?

The answer appears to be that in spite of the extensive improvements in all of these forms of mechanical technology, none are economically relevant in settings where the real value of human labor is extremely low. The difference between the real value of human time in economies like Indonesia or Bangladesh and the more advanced developing countries or the modern developed country are huge. In the lowest-wage economies, hand processes for almost all activities are the lowest-cost technologies. A century of intensive mechanical improvement activity has not yielded anything to change this.

We do observe that when wage rates rise, labor-saving machines

8. Paul A. David, *Technical Choice Innovation and Economic Growth*, (Cambridge: Cambridge University Press, 1975), pp. 197-200.

are not only adopted in low income countries, but are improved as well. It appears that for much of the mechanical technology in agriculture, there is scope for adoptive invention. This invention is important because, even though it does not produce major changes in machines, it produces improvements in "appropriateness."

Very few developing countries have as yet derived effective policies to encourage this adaptive invention. Some countries, such as India, rely heavily on public research investments. Others, such as the Philippines, encourage it by operating a "petty" patent or utility model patent system. The latter approach appears to be quite effective in stimulating this type of invention.

EVIDENCE ON RETURNS TO INVESTMENT IN PRODUCTION OF APPROPRIATE AGRICULTURAL TECHNOLOGY

This article has stressed the extent to which genotype-environment interactions and difference in geoclimate environments limit the diffusion and spread of biological technology. A similar interaction effect with the economic environment, chiefly the abundance or scarcity of labor and mechanical technology, has also been noted. It was further noted that there appeared to be scope in both biological and mechanical technology for technology developers to increase the adaptability of technology, and for recipients of this technology to adapt and modify it further to local conditions.

A summary of investment in agricultural research by developing countries showed that prior to 1950 the tropical developing countries of the world had significant research programs in place only for those commodities important in colonial trade along with a few small programs on rice and wheat. No significant work on root crops, oil seeds, pulses, sorghum, millets, and other feed grains was being undertaken. During the 1950s and 1960s a number of research institutions were built in the developing world, usually with international support.

National governments in the 1950s were not according high priority to the agricultural section—this was the period of import substitution policies to stimulate industrial growth—and certainly not to agricultural research. Even when international aid financed the training of agricultural scientists, many national governments failed to provide research facilities and other support. In response to this situation, the International Centers Research System was developed during the 1960s and 1970s. The centers were interdisciplinary and directed attention to a limited number of commodities. They took an international perspective and thus placed emphasis on wide adaptability.

We now have sufficient experience with the agricultural research systems in both the developed and developing countries to evaluate at least partly whether they have produced research products of value. It would be useful to know how the national programs in the developing countries have performed relative to the international centers and to systems in developed countries.

Table 3 provides a summary of a number of studies of agricultural research programs, including a number of specialized commodity programs. These studies use two basic methods for evaluation: impu-

TABLE 3
SUMMARY STUDIES OF AGRICULTURAL RESEARCH PRODUCTIVITY

STUDY	COUNTRY	COMMODITY	TIME PERIOD	ANNUAL INTERNAL RATE OF RETURN (percentage)
INDEX NUMBER				
Griliches, 1958	USA	Hybrid corn	1940-55	35-40
Griliches, 1958	USA	Hybrid sorghum	1940-57	20
Peterson, 1967	USA	Poultry	1915-60	21-25
Evenson, 1969	South Africa	Sugarcane	1945-62	40
Ardito Barletta, 1970	Mexico	Wheat	1943-63	90
Ardito Barletta, 1970	Mexico	Maize	1943-63	35
Ayer, 1970	Brazil	Cotton	1924-67	77+
Schmitz & Seckler, 1970	USA	Tomato harvester	1958-69	
		With no compensation to displaced workers		37-46
		Assuming compensation of displaced workers for 50 percent of earnings loss		16-28
Ayer & Schuh, 1972	Brazil	Cotton	1924-67	77-110
Hines, 1972	Peru	Maize	1954-67	35-40* 50-55†
Hayami & Akino, 1977	Japan	Rice	1915-50	25-27
Hayami & Akino, 1977	Japan	Rice	1930-61	73-75
Hertford, Ardila, Rocha & Trujillo, 1977	Colombia	Rice	1957-72	60-82
	Colombia	Soybeans	1960-71	79-96
	Colombia	Wheat	1953-73	11-12
	Colombia	Cotton	1953-72	None
Pee, 1977	Malaysia	Rubber	1932-73	24
Peterson & Fitzharris, 1977	USA	Aggregate	1937-42	50
			1947-52	51
			1957-62	49
			1957-72	34
Wennergren & Whitaker, 1977	Bolivia	Sheep	1966-75	44.1
		Wheat	1966-75	-47.5
Pray, 1978	Punjab (British India)	Agricultural research and extension	1906-56	34-44
	Punjab (Pakistan)	Agricultural research and extension	1948-63	23-37
Scobie & Posada, 1978	Bolivia	Rice	1957-64	79-96
PRODUCTION FUNCTION				
Tang, 1963	Japan	Aggregate	1880-1938	35
Griliches, 1964	USA	Aggregate	1949-59	35-40
Latimer, 1964	USA	Aggregate	1949-59	Not significant
Peterson, 1967	USA	Poultry	1915-60	21
Evenson, 1968	USA	Aggregate	1949-59	47
Evenson, 1969	South Africa	Sugarcane	1945-58	40
Ardito Barletta, 1970	Mexico	Crops	1943-63	45-93
Duncan, 1972	Australia	Pasture improvement	1948-69	58-68
Evenson & Jha, 1973	India	Aggregate	1953-71	40
Kahlon, Bal, Saxena & Jha, 1977	India	Aggregate	1960-61	63
Lu & Cline, 1977	USA	Aggregate	1938-48	30.5
			1949-59	27.5
			1959-69	25.5
			1969-72	23.5

TABLE 3 Continued

STUDY	COUNTRY	COMMODITY	TIME PERIOD	ANNUAL INTERNAL RATE OF RETURN (percentage)
Bredahl & Peterson, 1976	USA	Cash grains	1969	36‡
		Poultry	1969	37‡
		Dairy	1969	43‡
		Livestock	1969	47‡
Evenson & Flores, 1978	Asia— national	Rice	1950-65 1966-75	32-39 73-78
	Asia— international	Rice	1966-75	74-102
Flores, Evenson & Hayami, 1978	Tropics	Rice	1966-75	46-71
	Philippines	Rice	1966-75	75
Nagy & Furtan, 1978	Canada	Rapeseed	1960-75	95-110
Davis, 1979	USA	Aggregate	1949-59 1964-74	66-100 37
Evenson, 1979	USA	Aggregate	1868-1926	65
	USA	Technology oriented	1927-50	95
	USA—South	Tech. oriented	1948-71	93
	USA—North	Tech. oriented	1948-71	95
	USA—West	Tech. oriented	1948-71	45
	USA	Science oriented	1927-50 1948-71	110 45
	USA	Farm manage- ment research & agri- cultural extension	1948-71	110

*Returns to maize research only.

†Returns to maize research plus cultivation 'package.'

‡Lagged marginal product of 1969 research on output discounted for an estimated mean lag of 5 years for cash grains, 6 years for poultry and dairy, and 7 years for livestock.

SOURCES: The results of many of the studies reported in this table have previously been summarized in the following:

Thomas M. Arndt, Dana G. Dalrymple, and Vernon W. Ruttan, eds. *Resource Allocation and Productivity in National and International Agricultural Research* (Minneapolis: University of Minnesota Press, 1977), pp. 6, 7. • James K. Boyce and Robert E. Evenson, *Agricultural Research and Extension Systems* (New York: The Agricultural Development Council, 1975) p. 104. • Robert Evenson, Paul E. Waggoner, and Vernon W. Ruttan, "Economic Benefits from Research: An Example from Agriculture," *Science* 205 (14 Sept. 1979), pp. 1101-07. • Robert J.R. Sim and Richard Gardner, *A Review of Research and Extension Evaluation in Agriculture* (Moscow: University of Idaho, Department of Agricultural Economics Research Series 214, May 1978), pp. 41, 42. • R. Hertford, J. Ardila, A. Rocha, and G. Trujillo, "Productivity of Agricultural Research in Colombia," in *Resource Allocation and Productivity*, Thomas M. Arndt, Dana G. Dalrymple, and Vernon W. Ruttan, eds. (Minneapolis: University of Minnesota Press, 1977), pp. 86-123. • J. Hines, "The Utilization of Research for Development: Two Case Studies in Rural Modernization and Agriculture in Peru" (Ph.D. dissertation, Princeton University, 1972). • A. S. Kahlon, H. K. Bal, P. N. Saxena, and D. Jha, "Returns to Investment in Research in India," in *Resource Allocation and Productivity*. Thomas M. Arndt, Dana G. Dalrymple, and Vernon W. Ruttan, eds. (Minneapolis: University of Minnesota Press, 1977), pp. 124-47. • R. Latimer, "Some Economic Aspects of Agricultural Research and Extension in the U.S." (Ph.D. dissertation, Purdue University, 1964). • Y. Lu and P. L. Cline, "The Contribution of Research and Extension to Agricultural Productivity Growth," paper presented at summer meetings of American Agricultural Economics Association, San Diego, 1977. • J. G. Nagy and W. H. Furtan, "Economic Costs and Returns from Crop Development Research: The Case of Rapeseed Breeding in Canada," *Canadian Journal of Agricultural Economics* 26 (Feb. 1978), pp. 1-14. • T. Y. Pee, "Social Returns from Rubber Research on Peninsular Malaysia" (Ph.D. dissertation, Michigan State University, 1977). • W. L. Peterson, "Returns to Poultry Research in the United States," *Journal of Farm Economics* 49 (Aug. 1967), pp. 656-69. • W. L. Peterson and J. C. Fitzharris, "The Organisation and Productivity of the Federal-State Research System in the United States," in *Resource Allocation and Productivity*, Thomas M. Arndt, Dana G. Dalrymple, and Vernon W. Ruttan, eds. (Minneapolis: University of Minnesota Press, 1977), pp. 60-85. • C. E. Pray, "The Economics of Agricultural Research in British Punjab and Pakistani Punjab, 1905-1975" (Ph.D. dissertation, University of Pennsylvania, 1978). • A. Schmitz and D. Seckler, "Mechanized Agriculture and Social Welfare: The Case of the Tomato Harvester," *American Journal of Agricultural Economics* 52 (Nov. 1970), pp. 569-77. • G. M. Scobie and R T.

(Continued)

tation and statistical. Approximately half are based on developing country experiences. All report an estimated "internal rate of return" on investment. This computation treats research expenditures as an investment. The flow of increased commodity production, holding all inputs constant, is treated as the benefits stream. The internal rate of return is the rate realized over the entire period during which costs are incurred and benefits received. Some studies estimated the time lag between the time costs are incurred and benefits realized. The average estimated time lag between research spending and the full realization of its effect is roughly 10 years.

The imputation studies have each attempted to measure the costs and benefits to a particular program of research conducted over the time periods indicated. Different methods and data are utilized to measure the benefits. Sometimes statistical procedures were used, in other cases data comparing production using old and new technology were used. These studies report what might be termed average rates of return—that is, rates of returns that hold for the entire research investment.

The statistical studies, on the other hand, estimate a rate of return to an additional or marginal dollar of research spending. They generally employ an aggregate production function that is estimated utilizing data on production, inputs, and public sector programs such as research and extension. The research variables have to be specified carefully as to their timing and spatial dimensions. They are subject to the normal statistical bias for drawing inferences. Most of the studies reported "statistically significant" estimates of research effects.

Without discussing each study in detail, the following characterizations may be made.

—Only three studies report low rates of return. All others are in excess of 20 percent, in real terms.

—The imputation and statistical studies yield similar estimates.

—Estimates for research programs in developing countries are of roughly the same order of magnitude as those for more advanced countries.

(Table 3 notes continued)

Posada, "The Impact of Technical Change on Income Distribution: The Case of Rice in Colombia," American Journal of Agricultural Economics 60 (Feb. 1978), pp. 85-92. • A. Tang, "Research and Education in Japanese Agricultural Development," Economic Studies Quarterly 13 (Feb.-May 1963), pp. 27-41 and 91-9. • E. B. Wennergren and M. D. Whitaker, "Social Return to U.S. Technical Assistance in Bolivian Agriculture: The Case of Sheep and Wheat," American Journal of Agricultural Economics 59 (Aug. 1977), pp. 565-69.

The sources for the individual studies are as follows:

H. Ayer, "The Costs, Returns and Effects of Agricultural Research in Sao Paulo, Brazil" (Ph.D. dissertation, Purdue University, 1970). • H. W. Ayer and G. E. Schuh, "Social Rates of Return and Other Aspects of Agricultural Research: The Case of Cotton Research in Sao Paulo, Brazil," American Journal of Agricultural Economics 54 (Nov. 1972), pp. 557-69. • N. Ardito Barletta, "Costs and Social Benefits of Agricultural Research in Mexico" (Ph.D. dissertation, University of Chicago, 1970). • M. Bredahl and W. Peterson, "The Productivity and Allocation of Research: U.S. Agricultural Experiment Stations," American Journal of Agricultural Economics 58 (Nov. 1976), pp. 684-92. • R. C. Duncan, "Evaluating Returns to Research in Pasture Improvement," Australian Journal of Agricultural Economics 16 (Dec. 1972), pp. 153-68. • R. Evenson, "The Contribution of Agricultural Research and Extension to Agricultural Production" (Ph.D. dissertation, University of Chicago, 1968). • R. Evenson, "International Transmission of Technology in Sugarcane Production" (Yale University, New Haven, 1969). (Mimeographed.) • R. E. Evenson and D. Jha, "The Contribution of Agricultural Research Systems to Agricultural Production in India," Indian Journal of Agricultural Economics 28 (1973), pp. 212-30. • Z. Griliches, "Research Costs and Social Returns: Hybrid Corn and Related Innovations," Journal of Political Economy 66

(Continued)

—The estimate for international rice research is one of the highest reported. A similar estimate for wheat research at CIMMYT, while not made, would be similar. However, other centers have not produced results of this type.

These studies are open to criticism on a number of points, but even taking criticism into account, the general results hold. We are left, then, with the conclusion that policies toward investment in the production of appropriate agricultural technology are far from optimal. The fact that most such investment probably has to take place in the public sector is important to understanding why this is so. Public sector policymakers, often taking cues from international advisors, have persistently overestimated the extent to which appropriate technology will "spill in" to the sector and can be had at low cost. The agricultural sector in many developing countries, until recently, has not had high priority in national plans. The most successful agricultural research systems have a local clientele of farmers who support them and alter their programs. The political mechanisms for such support systems are not established in much of the world.

Nonetheless, the picture is not wholly pessimistic. It seems clear that with a few exceptions, developing countries are moving in the direction of more optimal investment. The long and difficult process of building research institutions is proceeding in many countries. The international centers—at least some of them—are filling in many gaps in the system and are channeling valuable genetic and other raw materials to these programs. In some countries invention is being encouraged. The result of this progress toward a more optimal investment program is revealing itself in improved productivity growth in many agricultural sectors.

(Table 3 notes continued)
(1958), pp 419-31.   • Z. Griliches, "Research Expenditures, Education and the Aggregate Agricultural Production Function," American Economic Review 54 (Dec. 1964), pp. 961-74.   • Y. Hayami and M. Akimo. "Organisation and Productivity of Agricultural Research Systems in Japan," in Resource Allocations and Productivity, Thomas M. Arndt, Dana G. Dalrymple, and Vernon W. Ruttan, eds. (Minneapolis: University of Minnesota Press, 1977), pp. 29-59.

ANNALS, *AAPSS*, 458, November 1981

The Green Revolution as a Case Study in Transfer of Technology

By CARL E. PRAY

ABSTRACT: The green revolution was a transfer of the idea of fertilizer-responsive grain varieties and the capacity to develop them from temporate countries to the countries of South and Southeast Asia, the Middle East, and Latin America. Key actors in this transfer of technology included the public research institutions in the less-developed countries (LDCs), the International Agricultural Research Centers, the Ford and Rockefeller Foundations, and the United States Agency for International Development. Once these institutions had bridged the gap between countries, the farmers rapidly accepted the new technology in areas where the agroclimatic and economic conditions were favorable. The green revolution was neither the cure for the problems of world hunger, as some early enthusiasts suggested, nor an important cause of income inequality and poverty, as suggested by its critics. When separated from the impact of factors such as rapid population growth, the shortage of arable land, and government policy, it is clear that the green revolution has substantially increased the supply of food grain and thus kept food grain prices lower than they would have been in the absence of new technology. This lowering of prices generally had a positive impact on income distribution. At the same time these varieties have had a less positive impact on agricultural income through their impact on demand for factors of production: landowners have benefited more than laborers. However, laborers would have been in a worse position in the absence of the green revolution.

Carl E. Pray is currently a research associate, Department of Agricultural and Applied Economics, University of Minnesota. He was the Bangladesh Associate of the Agricultural Development Council for two years. In Bangladesh he worked as an economist with the Bangladesh Agricultural Research Council and also participated in research projects with the Bangladesh Ministry of Agriculture and Planning Commission. His Ph.D. dissertation in economic history at University of Pennsylvania was on the economics of agricultural research and extension in Pakistan. He worked with the Peace Corps for two years as agricultural extension agent in India.

I T is now over 15 years since the new wheat and rice varieties were introduced on a wide scale in Asia, the Middle East and Latin America.[1] They have not lived up to the original overenthusiastic promise of some of their original promoters: they did not end world hunger; they did not lead to food self-sufficiency in Third World countries. At the same time the green revolution has not fulfilled the predictions of its critics: it did not cause widespread unemployment; neither farm size nor tenure have been important determinants of the utilization of the new technology; and it has not caused a rapid decline in the environment of less-developed countries (LDCs).

What it is responsible for is something between the extravagant promises of the early promoters and the gloom of the critics. It has led to a large increase in the production of wheat and rice in Asia, Latin America, the Near East, and North Africa. This increase has meant increased income to many farmers and landowners and increased employment for many landless laborers. However, regional dispar-

1. This article draws heavily on previous reviews of the green revolution. The following were particularly useful: Part IV of Yujiro Hayami and Vernon Ruttan, *Agricultural Development: An International Perspective* (Baltimore: Johns Hopkins University Press, 1971); Vernon W. Ruttan and Hans Binswanger, "Induced Innovation and the Green Revolution" in H. Binswanger and V. W. Ruttan et al., *Induced Innovation: Technology, Institutions, and Development* (Baltimore: Johns Hopkins University Press, 1978); Dana G. Dalrymple, *Development and Spread of High-Yielding Varieties of Wheat and Rice in the Less Developed Nations*, USDA, Foreign Agricultural Economic Report No. 95, 6th ed. (Washington, DC: Government Printing Office, Sept. 1978); and Michael Lipton, "Inter-Farm, Inter-Regional and Farm-Non-Farm Income Distribution: The Impact of the New Cereal Varieties," *World Development*, 6(3):319-37 (1978).

ities within the LDCs have frequently widened, and within the regions that gained, all may have benefited, but landlords benefited more than laborers. In many of the countries there is evidence that the benefits from increased production were passed along to consumers in the form of lower food prices. This tends to help the poor more than the rich consumer, as the poor spend a far greater percentage of their income on food. A final and somewhat unexpected beneficiary of this research has been the United States. Its wheat and rice farmers have benefited by using varieties based on the high-yielding varieties (HYVs) developed in Mexico and the Philippines. Consumers have benefited from lower prices because foreign demand for U.S. farm products was less than it would have been in the absence of new varieties.

This article has four sections. The first describes the nature of the transfer of technology. Then the factors in developing nations that made this transfer a success will be examined. The third section covers the impact of the HYVs on different groups in LDC societies. Finally, I will draw together the major lessons of the green revolution.

WHAT WAS ACTUALLY TRANSFERRED?

The essence of the green revolution was the transfer of the highly fertilizer-responsive grain varieties and the capacity to produce such varieties from Japan, Taiwan, the United States, and Europe to the developing countries of Asia, Latin America, the Middle East, and Africa. This type of biological innovation became an economically viable proposition in most developing countries only after World War II. At that time rapidly increasing population and the reduced availability of land in many countries led to increasing prices of foodgrains. There was a decline in chemical fertilizer prices due to technological change in the Western fertilizer industry in the 1950s. Thus fertilizer-responsive varieties became economical in countries where they previously had not been due to low foodgrain price/fertilizer price ratios.

Several attempts were made to transfer this technology in the 1950s and early 1960s. Varieties of wheat and rice from temperate zones were introduced into the breeding programs of many countries through bilateral exchange and aid agreements. The Food and Agriculture Organization of the United Nations coordinated wheat and rice breeding schemes. The varieties with the dwarfing gene that went into the HYV rices were used in some of these programs. They were not able to produce the needed breakthrough, although the Philippine, Indian, and Malaysian research programs seem to have produced fertilizer-responsive varieties or were getting close.[2]

The breakthrough in the development of varieties that were high yielding with increased use of fertilizer and that led to the green revolution was due to the International Agricultural Research Centers and the Mexican predecessor of the International Maize and Wheat Improvement Center (CIMMYT). The key breakthrough was the development of short-statured,

2. Useful reviews of the early rice and wheat research are: N. Parthasarathy, "Rice Breeding in Tropical Asia up to 1960," in *Rice Breeding* (Los Banos: IRRI, 1977); and E. C. Stakman, Richard Bradfield, and Paul C. Manrgelsdorff, *Campaigns Against Hunger* (Cambridge: Harvard University Press, 1967).

fertilizer-responsive, photoinsensitive rice and wheat varieties. These varieties could be used as model plant types for breeders to adapt to local conditions. They provided the basic genetic material to be used in breeders' development programs. Also, they produced spectacular increases in yield under ideal conditions, which convinced governments of the benefits of the HYV programs. The centers' second contribution was their formal and informal extension network of returned trainees from their training programs, Rockefeller and Ford Foundation personnel and International Rice Research Institute (IRRI) and CIMMYT representatives, who were all making the point that substantial gains in output could be made through HYVs and fertilizer inputs. Their third main contribution was improved research methodology and better organization, which increased the efficiency of local research.

The first users of the new wheat and rice varieties were India, Malaysia, and Pakistan in 1965. Since then Dalrymple has recorded the spread of wheat varieties to over 30 countries and the rice varieties to almost 50 countries. This spread includes not only the market-oriented LDCs but also the People's Republic of China and Cuba. The position of these varieties in 1976-1977 is shown in Table 1.

Hayami and Ruttan have identified three stages of technology transfer.[3] Material transfer "is characterized by the simple transfer or import of new materials such as seed, plants, animals, machines, and techniques associated with these materials. Local adaptation is not conducted in an orderly and systematic fashion." In

3. Hayami and Ruttan, p. 175.

TABLE 1
ESTIMATED AREA OF HIGH-YIELDING VARIETIES OF WHEAT AND RICE AND
PROPORTION OF CROP AREA PLANTED TO HYVS, LESS-DEVELOPED NATIONS,
1976-77*

REGION	WHEAT	RICE	TOTAL
HYV AREA		HECTARES	
Asia, South and East	19,672,300	24,199,900	43,872,200
Near East, West Asia and North Africa†	4,400,000	40,000	4,440,000
Africa (excl. N. Africa)†	225,000	115,000	340,000
Latin America	5,100,000	920,000	6,020,000
Total	29,397,300	25,274,900	54,672,200
HYV AREA/TOTAL AREA		PERCENTAGE	
Asia	72.4	30.4	41.1
Near East	17.0	3.6	16.5
Africa	22.5	2.7	6.5
Latin America	41.0	13.0	30.8
Total	44.2	27.5	34.5

SOURCE: Dana G. Dalrymple, "Development and Spread of High-Yielding Varieties of Wheat and Rice in the Less Developed Nations," USDA, Office of International Cooperation and Development, Foreign Agricultural Economic Report 95, 6th ed. (Washington, DC: Sept. 1978), pp. 122, 123.
*Excluding Communist nations, Taiwan, Israel, and South Africa.
†Particularly rough estimate of area.

the design transfer stage the transfer of technology is "primarily through the transfer of certain designs. . . . During this period the imports of exotic plant materials and foreign equipment are made in order to obtain new plant breeding materials or to copy equipment, rather than to use in direct production." Their final stage is capacity transfer, in which the transfer of technology is made "through the transfer of scientific knowledge and capacity which enable the production of locally adaptable technology, following the 'prototype' technology which exists abroad." Instead of bringing in prototype plants and machines to copy, new scientific techniques or advances in agricultural technology move through professional journals, newsletters, and, perhaps most importantly, through the movement of scientists.

The type of transfer that occurred during the green revolution depended on the agroclimatic similarity of the adopting country with the country of origin and also on the sophistication of the local research system. The initial diffusion of HYVs in India, Pakistan, Turkey, and Malaysia was largely a material transfer because their agroclimatic conditions were similar to Mexico and the Philippines. The governments of these countries imported large amounts of seed from Mexico, the Philippines, and Taiwan that were then either distributed directly to the farmers or were multiplied and then distributed. Some of these countries were at the same time working on the "design" and moving into the "capacity" stage. India and Malaysia started to develop their own wheat and rice varieties based on the dwarf prototypes at the same time that the IRRI and CIMMYT varieties were being pushed out to farmers. In other countries the spread of HYVs had to wait until the design and capacity stages had been reached so that HYVs could be designed to fit the economic or agroclimatic conditions in the country. Thus Thailand has had to wait for the local development of varieties that meet international trade standards, and Bangladesh has had to wait for the development of varieties for upland and lowland unirrigated conditions. In most of the countries that accepted HYVs there has been the development of research facilities so that countries could move to more advanced stages of technology transfer. At present the main method of transferring crop technology in Asia is through capacity transfer. This movement to the capacity stage has been an integral part of the green revolution.

WHAT DETERMINED THE ACCEPTANCE OF THESE NEW VARIETIES?

The question of what determined the acceptance of these new varieties has to be broken down into two parts. First is the question of why the government of a country decided to initiate a program of importing and/or developing HYV technology. Second is the question of the internal diffusion of HYVs once they were brought in or developed in a country. The first question focuses on government decision making while the second concentrates on the decision making of the farmers.

There are a number of factors that entered into the government's decision on HYVs. Their decision was based on two general factors: the size of the expected payoff and availability of information about that payoff. In many of the LDCs the price and availability of the main foodgrain can determine whether a government survives or not. Thus the payoffs from suitable HYVs to the government, which has low foodgrain prices as a goal, can be considerable.

However, the possible payoff to governments from HYVs depends not only on the size of the country's food deficit and its determination to reach self-sufficiency, but also on the suitability of the HYV's to the agroclimatic and economic conditions and the importance of wheat and rice in the country.

Furthermore, if the government is not aware of the potential of HYVs under the country's agroclimatic conditions, it will not import the prototypes or back the necessary investments in agricultural research. Thus the information flow from the outside through International Agricultural Research Centers (IARCs), the international press, academic journals, the movement of scientists, and the activities of foundations like Ford and Rockefeller becomes important. Equally important is the ability to evaluate this information, which is the role of the local scientific establishments.

No one, to my knowledge, has tried to test the relative importance of the various reasons for governments' decisions to spread the new varieties. The best that can be done at the moment is to point out examples and evidence that indicate that the factors were important. The food crises of the mid-1960s caused by drought plus, in the case of Pakistan, the cut-off of American food aid at the time of the 1965 Indian-Pakistani war were key factors in convincing India and Pakistan to import substantial amounts of HYV wheat seed from Mexico in 1967 and 1968.[4] They also imported rice seed from IRRI and Taiwan although not in such large amounts. Countries in

4. See Carl E. Pray, "Agricultural Research in British and Pakistani Punjab, An Induced Innovation Interpretation," Staff Paper, Dept. of Agricultural and Applied Economics, University of Minnesota, 1981.

which these crops were not important were not particularly interested in the HYVs. Thus the government of Bangladesh, where wheat is a minor crop, did not import or produce large quantities of HYV wheat seed until 1974, when the success of Mexican wheat in other countries became evident.

One of the most important factors in determining which governments participated in the green revolution first was the strength of the local research program. Strong research programs, such as the Indian Agricultural Research Institute, were able to spot the potential of the CIMMYT wheat varieties in the early 1960s. They were then able to push their government into importing sufficient amounts of seed to start the testing and then extension of the HYVs.[5] Finally, they were quickly able to use the new goals and methodologies developed at the international centers to develop their own varieties. Kislev and Evenson, in a survey of agricultural research and productivity, have been able to show statistically that research conducted in a second country with similar agroclimatic conditions will have a much greater impact on the first if the first has an agricultural research capacity of its own.[6] In an assessment of HYVs in Asia, Evenson concludes that "countries without significant indigenous research capability realized almost no transfer benefits."[7]

In addition to developing the prototype varieties the interna-

tional centers—CIMMYT and IRRI—played an important role in transmitting information about the new technology to the LDCs. They did this in a number of different ways. They held training courses and seminars for scientists and public officials. They pushed written material and publicity through the news media, academic journals, and their own publications. Scientists traveled throughout the LDCs publicizing the new varieties to governments and scientists. They frequently had resident advisors in LDCs. Finally, they set up an extensive network for testing and exchanging genetic material.

In South Asia, the Ford Foundation and the Rockefeller Foundation played quite important roles in the green revolution by providing the initial financial support for the IARCs. Also, the local offices of the foundations in South Asia played an important role in convincing governments to try these new varieties and then financed some of the early imports of HYV seed.

However, it is interesting to note that the United States government did not play a direct role in developing HYV wheat and rice varieties until 1969, when it began to support CIMMYT. The 1977 World Food and Nutrition Survey of the National Research Council of the United States reports as follows:

An AID official acknowledged in February 1966: "We have not focused research attention on the increase of production of crops such as rice and wheat, which have been in surplus in the United States. This reflected the attitude of the Congress, of the American public, and of American farm organizations—a handicap that is still to be overcome."[8]

5. Dalrymple, p. 16.

6. Robert Evenson and Yoav Kislev, "Research and Productivity in Wheat and Maize," *Journal of Political Economy*, 81 (Sept.-Oct. 1973).

7. Robert E. Evenson, "Agricultural Research and Extension in Asia: A Survey," in *Rural Asia: Challenge and Opportunity. Second Asian Agricultural Survey Supplementary papers Vol. II* (Manila: Asian Development Bank, 1978), p. 33.

8. World Food and Nutrition Study, Volume V, Study Team 14, Agricultural Research Organization (Washington, DC: National Academy of Sciences, 1977), p. 95.

However, since 1969 the U.S. Agency for International Development (USAID) has invested an increasingly large share of their budget into the international centers and national research programs, and at present it makes the largest single monetary contribution to the IARCs.

In several Asian countries USAID's institution-building programs were important in speeding the spread of HYVs. For example, the agricultural universities in India played a key role in spreading HYV wheat, and the Indian Fertilizer Association was important in popularizing chemical fertilizer with farmers and in developing the local fertilizer industry. USAID played a major role in the development of these institutions. Also, AID provided the Indian government with Mexican wheat seed.

In Ruttan and Binswanger's review of the literature on the green revolution two generalizations about farmers' acceptance of the new varieties were supported by the empirical evidence: First, "the new wheat and rice varieties were adopted at exceptionally rapid rates in those areas where they were technically and economically superior to local varieties." They point out the examples of Indian and Pakistani provinces that went from negligible amounts of HYVs in 1965-1966 to about 70 percent of the wheat acreage in 1969-1970.[9] Similar results were found by Barker and Pal in 1979 for rice areas that were appropriate for the growing of HYV rice.[10] These findings should have laid to rest the belief in the irrationality of Asian farmers once and for all. The second generalization is that

9. Ruttan and Binswanger, p. 386.
10. Randolph Barker and T.K. Pal, "Barriers to Increased Rice Production in Eastern India," IRRI Research Paper Series, No. 25 (Los Banos: IRRI, Mar. 1979).

neither farm size nor farm tenure has been a serious constraint on the adoption of new high-yielding varieties. Differential rates of adoption by farm size and tenure have, of course, been observed. The available data seem to imply, however, that within a relatively few years after introduction lags in adoption rates owing to size or tenure have typically disappeared.[11]

IMPACT OF THE GREEN REVOLUTION ON SOCIETY

Government officials in both the LDC governments and some of the aid agencies, many of the journalists, and even some of the scientists—although most of them were more cautious—promised far more than the green revolution varieties could deliver. What I will try to do here is to compare the accomplishments and problems of the green revolution, not with the early inflated promises, but with the situation that would have existed in the absence of the green revolution. By analyzing the green revolution this way we should be able to sort out what has happened in conjunction with, but not been caused by, the green revolution and so arrive at a more accurate assessment of its impact.

How much more grain is now produced than would have been produced without the green revolution? To answer this question we have to make several important assumptions about what the growth in grain production would have been. Thus we should compare the actual observed growth rate in grain production with the growth rate generated by an assumed research program and an assumed growth of output prices. No one has attempted to develop such an estimate. Even-

11. Ruttan and Binswanger, p. 386-9.

son and Flores have come the closest.[12] They have divided the actual gains in Asian rice production into the part due to HYVs, the part due to local research, and the part due to an interaction of the HYVs and local research. They concluded that the increase in output due to HYV rice in all LDCs amounted to between 7 and 12 percent of the output of rice in 1974-75. This represents a sizable shift in the supply curve of foodgrains. In Asia about half of this was due to directly transferred IRRI varieties, 25 percent due to joint IRRI-national varieties, and the rest due to locally developed HYVs. Therefore, only 75 percent of the increased production was related to the green revolution. Using a less sophisticated methodology, Andersen[13] has estimated the increase in world output due to modern rice varieties to be 10 million tons, or 5.4 percent of the total rice production, and 21 million tons, or 22.4 percent of wheat production, due to new wheat varieties.

Increases in production of this size clearly have had an effect on the world market of wheat and rice. In aggregate this increased availability has led to world prices that are lower than they would have been in the absence of HYVs. This clearly benefits the importing governments and some poor consumers who now have to pay less for their basic food. This effect has been frequently ignored because the increases in population and income have increased demand more rapidly than supply, with the result that

prices of grain have continued to increase. In addition to the aggregate impact, there have been changes in trading patterns, at least periodically, as certain countries have become self-sufficient in food-grains. These include the Philippines around 1970, South Korea, and India in the last few years.

In closed economies with no government intervention in grain markets, increases in food production led to lower prices for food. Lower food prices are beneficial to all consumers who have to buy their food in the market rather than produce it themselves. This means the urban population and rural non-farm families and agricultural laborers. Since these groups included the poorest members of society, and since the poorest members of society spend most of their income on food—up to 70 or 80 percent in many countries in Asia—this drop in prices has substantial positive effects on the income distribution in these countries. In general, the gain in output has either kept prices lower, reduced government expenditures on foodgrain subsidies, and/or improved the country's foreign exchange position.

Despite the early criticism that the green revolution led to more unemployment, there now seems to be a consensus among scholars about the green revolution's impact on labor.[14] First, the HYVs have increased the demand for labor in Asia even in areas where there has been considerable mechanization. Second, the demand for labor has not, however, been growing as fast as the supply of labor, and thus the real income of the agricultural

12. Robert Evenson and P.M. Flores, "Social Returns to Rice Research," in *Economic Consequence of the New Rice Technology* (Los Banos: IRRI, 1978).

13. Per Pinstrup-Andersen, *The Role of Agricultural Research and Technology in Economic Development* (New York: Longman, forthcoming), Table 5-6.

14. See Keith Griffin, "Comments on Labor Utilization in Rice Production," in *Economic Consequences of the New Rice Technology.*

laborer in many South and South-east Asian countries has been steadily declining over the last 20 years. Third, mechanized cultivation in Asia had started before the green revolution and has not been noticeably speeded up by the introduction of HYVs.

The impact of mechanized cultivation has been to decrease the demand for labor without increasing productivity, and its spread has been speeded by government policies that are independent of the green revolution. The other type of mechanization that appears to be more closely associated with the green revolution is mechanized irrigation in the form of power pumps and tubewells. The impact of this type of mechanization has been to increase the demand for labor because it has allowed an increase in multiple cropping.

Much of the early criticism of the HYVs regarding labor seems to have arisen because critics confused independent trends—specifically, growth in rural population and mechanized cultivation—with the effect of HYVs. A number of recent studies have analyzed the impact of the HYVs on labor utilization. They show that the demand for labor increased but not enough to keep up with population growth.[15] Thus the condition of the landless laborer is better than it would have been in the absence of HYVs, but his position is declining and the HYVs by themselves cannot reverse this situation.

The criticism that has been leveled at the green revolution most

consistently is that it has worsened the rural income distribution. There seems to be a consensus that landowners have captured more of the gains from the HYVs than the tenants and laborers. A study of the distribution of the increased output from HYVs in the Aligarh district of India showed 67 percent went to owners of land and capital, 23 percent to sellers of inputs like fertilizer, and 10 percent to laborers. This seems to have been rather typical of all India.[16] This does not seem to be due to the biased shift in technology. Relative to the other important technological path for improving agriculture—mechanization—HYVs are clearly land-saving if they are biased at all. The reasons that this neutral shift in technology has led to a large increase in the rents to land is that the supply of labor is relatively elastic compared with the supply of land and, second, that the supply of labor is shifting out at a very rapid rate. In the absence of HYVs, pressure on land would have been even higher and income distribution more skewed toward landowners.

Among landholders we again have a situation where technology is neutral, but existing structural inequities and independent shifts in supply of factors of production have led to an unequal distribution of the gains. There has been no evidence that there are economies of scale in this technology that give big farmers an advantage over small farmers in its use; and despite years of searching, there is almost no evidence that HYVs have been adopted more slowly on sharecropped or leased land than on owner-operated land. Even some of the critics agree that it is not the technology that has

15. For example, Randolf Barker and B. G. Cordova, "Labor Utilization in Rice Production," in *Economic Consequences of the New Rice Technology.*

16. Ruttan and Binswanger, p. 390.

led to unequal distribution of gains; rather, it is the access to inputs that are rationed by political rather than economic processes that lead to the inequality.[17] Lipton's extensive survey of the literature finds little evidence that there are economies of scale in the use of the new varieties, but he finds "scale economies in product distribution and storage, and in obtaining inputs, are unquestioned, as are higher per-acre costs of administration and extension for small farmers."[18] HYVs did not cause this situation, and they cannot solve it.

The environmental impact of the green revolution is one of the least researched facets of the change. It seems safe to say that on balance, the new technology has had a positive impact on the land base of agriculture. Prinstrep-Anderson says the following:

Technological change resulting in higher yields reduces the pressures for expanding agricultural production into marginal lands and excessive exploitation of the land base. There is little doubt that the development and use of high yielding crop varieties, expanded use of fertilizers, better production practices and other yield increasing factors have been of great importance in limiting land degradation in developing countries.[19]

Negative effects of the HYVs could come through increased use of fertilizer, pesticides, and water, which could lead to declining quality of water or soil through chemical pollution or salinization, but the evidence is not conclusive.

Another problem associated with the spread of HYVs is the decrease in genetic diversity in areas where HYVs have been widely distributed.[20] This causes two kinds of problems. First, the uniformity of genetic background of HYVs means that there is the possibility that a new disease or pest to which they are not resistant could sweep through an entire area, causing a large crop failure.

The second problem is that the reserves of genetic diversity that allow breeders to produce varieties resistant to new diseases and other stresses are being lost as farmers in the Third World, who grew many varieties and thus were a major source of genetic variability, switch to HYVs. Much is being done to preserve these resources, and a major role is being played by the International Board for Plant Genetic Resources, which was created in 1973, and the International Agricultural Research Centers. It is too early to say whether enough is being done.

The impact of the green revolution that we have discussed up to this point has been the direct impact on the increased production of grain on individuals and households in society. However, the increased food production and income due to the new technology also led to changes in the structure of society and to changes in government policy. Critics have suggested that the increased production has led to a breakdown in traditional village structures that acted to even out the impact of disasters such as famine—sharecropping and paying harvesters with a share of the harvest—and to a worsening of the distribution of income and power with the richer members of the community gaining control over land. The problem again is whether these changes can be attributed to the green revolution

17. Griffin.
18. Lipton, p. 325.
19. Prinstrep-Andersen, p. 267.

20. This paragraph is drawn mainly from Andersen's Chapter 8.

or to other independent factors, such as the increases in population and decreased availability of land, which were taking place at the same time as the green revolution.

The changes in government policy due to the green revolution have received little attention from researchers. Some of the critics are disturbed because it has distracted the attention of the governments and people of the LDCs and DCs from what they view as the real prerequisite of sustained growth: more equal distribution of power and income. The implication is that in the absence of the green revolution, the social changes that are required for more equality, such as land reform, would have taken place more rapidly in its absence. This is a proposition that could be empirically tested. Did effective land reform take place more often in countries unaffected by the green revolution? Are poorer populations more likely to change the social structure than somewhat more wealthy and healthy ones?

The land reforms in Yugoslavia in the 1920s, Japan in the late 1940s, Taiwan in the early 1950s, and Iran in the 1960s took place when rapid technological change was already taking place.[21] Also, the new technology in a number of countries has led to discussion and research on the constraints that have prevented its spread. HYVs enabled people to measure the importance of constraints, such as the availability of inputs and the land tenure system, and to put pressure on governments to correct these constraints. The absence of high-yielding varieties in a particular area suited to them is highly visible evidence that something is wrong. The connections between poverty or technological

change and land reform or other basic measures to increase equality of income and power are still not clearly understood. More research is required.

Other shifts in government activity due to the green revolution have been documented. The first is the shift of emphasis that has taken place in the research establishments of Asian countries away from cash crops into food crops. This shift has been documented in Pakistan.[22] There, the availability of HYVs induced a shift of resources into wheat research in the mid-sixties, and then rapid success of wheat led to increased expenditure in other food crops.

It needs to be stated clearly that the benefits from this transfer of technology have not been restricted to the developing countries. The U.S. consumer has benefited from food prices that are somewhat lower than they would have been without the green revolution because there has been less foreign demand for U.S. wheat and rice, and also because dwarf varieties derived from green revolution varieties are now increasing food production in the United States. Dana Dalrymple has traced the spread of semidwarf varieties in the United States; his data show that a small area of rice and, by 1974, almost four million acres of wheat in the United States were planted with varieties descended from material developed in Mexico, and by 1979 it was much higher than that.[23]

LESSONS FROM THE GREEN REVOLUTION

The goal of the LDCs is at least to develop the local technical and

21. Hayami and Ruttan, p. 262.

22. Pray.

23. Dana Dalrymple, *Development and Spread of Semi-Dwarf Varieties of Wheat and Rice in the United States: An International Perspective* (Washington: USDA, 1980).

scientific capacity that can make the decisions about what is appropriate technology for meeting the goals of the developing society, and that in the longer run can develop the needed technology themselves. The transfer of specific technologies then becomes a short-run substitute for the local capacity and also can be used as part of a program to help develop local capacity. If we in the West are going to assist the process of meeting this goal, we must learn from the successful transfers.

There are five major lessons from the green revolution. First, it is clear that for a technology to be transferred, the research and technology has to be in some sense appropriate to both the physical and economic conditions in the country where it is being transferred. Thus high-yielding varieties had to be developed that would yield more than local varieties under local agroclimatic conditions with the new economic conditions of low fertilizer prices. In contrast, earlier attempts to shift American or British wheat varieties directly to the tropics during the colonial period failed to produce increased yields, as did later local breeding programs that did not adjust to the new economic conditions of low fertilizer prices.

Second, it is not sufficient simply to transfer a new variety or design. Without active local research programs to develop new fertilizer-responsive varieties in most countries, HYVs would not have spread as widely and also would have been subject to faster depreciation as new insects and diseases attacked them. Thus the capacity to produce new technology and to do research to maintain the technology is a necessary part of technology transfer.

Third, a more detailed history of the countries involved would probably suggest that the capacity for developing new technology can be more effectively developed if it is combined with a new technology such as the HYVs. For example, the USAID projects to develop agricultural universities in India in the early 1960s undoubtedly would have produced results that would have led to public support for these programs. However, research takes time to produce results, and in that time the public can lose interest and funding can decline. However, before this happened the green revolution varieties were introduced and several of these universities—particularly in Punjab and Uttar Pradesh—were able to take partial credit for these advances. This greatly increased their public support at a time when indigenous research breakthroughs were not yet ready. Thus the pattern of the green revolution period where a high-profile technology was coupled with capacity development seems to be an ideal way of speeding the transfer of technology, but it is hard to duplicate. The other method of simply developing the institutional capacity without being centered on a specific technology or scientific breakthrough would be slower to develop local support and less likely to succeed in the long run.

Fourth, external agencies such as the IARCs, the Ford and Rockefeller Foundations, and the developed country governments have played an important role in this transfer of agricultural technology. The foundations and the U.S. government have supported international centers and aided in the development of local scientific capacity through institutional support and the training of scientists. The government should continue to do so particularly today, when some major sources of funding, like the Ford Foundation, are withdrawing from this area.

Fifth, the impact of the technological change on society is determined by existing socioeconomic structure and the government policies, which in part reflect that socioeconomic structure. Even though the green revolution technology was itself neutral with regard to the size of the holding, the socioeconomic structure distributed most of the benefits to the larger farmers in many countries. Likewise, the distribution of benefits between farmers and consumers was determined by government pricing policies rather than the nature of the technology or unregulated supply and demand.

If in the short run we are involved in the development and transfer of some agricultural technology, we have the responsibility to choose technologies that are economically viable in the recipient country and do not worsen the major problems of the LDCs such as hunger, unemployment and underemployment, and degradation of the environment. A final condition is that we should choose programs that have the maximum value in helping to develop local scientific capacity.

Technology that does not pass the criterion of economic viability will not be used, and so pushing uneconomic technology will, at best, waste scientists' and technicians' time and, at worst, will decrease the prestige of the local research institute and the public's and government's appreciation of the value of science in general. Thus this criterion has to be adhered to, and then the other conditions should be checked. The other conditions mean in practice that people dealing with agricultural technology choice in Asia should be skeptical of mechanical cultivation that largely displaces agricultural laborers and pesticides that in the long run may not be economical and, as currently used, are dangerous to man and the environment. This does not mean that we should reject all mechanical devices or pesticides, but we should more carefully examine their impact before pushing them on developing nations. At the same time we should not ask too much of those who develop technology. They cannot be the main agents to achieve social change. As Vernon Ruttan points out, technical change is a blunt instrument for reform: "A nation's agricultural research budget can be a powerful instrument for expanding its capacity to produce food and fiber. It is a relatively weak instrument for changing income distribution in rural areas."[24]

24. Vernon W. Ruttan, *Agricultural Research Policy* (St. Paul: University of Minnesota Press, forthcoming).

The Market for Know-How and the Efficient International Transfer of Technology

By DAVID J. TEECE

ABSTRACT: This article explores the nature of international technology transfer and the operation of the market for know-how. It begins by examining the relationship between codification and transfer costs and then analyzes various imperfections in the market for know-how. The special properties of know-how are shown to confound various aspects of the exchange process when arms-length contracting is involved. The internalization of the exchange process within multinational firms serves to bypass many of these difficulties, and explains why the multinational firm is of such importance. Several forms of regulation of technology imports and exports are examined. It is discovered that the process is insufficiently well understood to permit the design of effective regulation that, moreover, appears unlikely to eliminate inefficiency. An efficiency focus is maintained throughout since I feel no qualification to pontificate on complex and confused distributional issues.

David J. Teece is associate professor of business economics at the Graduate School of Business, Stanford University. He has a bachelor's and master's degree from the University of Canterbury, Christchurch, New Zealand, and a doctorate in economics from the University of Pennsylvania, which he received in 1975. Professor Teece specializes in the fields of industrial organization and the economics of technological change, and has published numerous scholarly articles and monographs on the multinational firm, the organization of the petroleum industry, the behavior of the Organization of Petroleum Exporting Countries (OPEC), and the relationship between the internal organization and performance of large enterprises.

NOTE: The financial support of the National Science Foundation is gratefully acknowledged, together with the valuable comments from Max Boisot, Almarin Phillips, and Oliver Williamson.

ECONOMIC prosperity rests upon knowledge and its useful application. International, inter-regional, and interpersonal differences in levels of living can be explained, at least in part, by differences in the production techniques employed. Throughout history, advances in knowledge have not been uniformly distributed across nations and peoples, but have been concentrated in particular nations at particular times. According to Kuznets,

. . . the increase in the stock of useful knowledge and the extension of its application are of the essence of modern economic growth. . . . No matter where these technological and social innovations emerge . . . the economic growth of any given nation depends on their adoption. In that sense, whatever the national affiliation of resources used, any single nation's economic growth has its base somewhere outside its boundaries—with the single exception of the pioneering nations.[1]

The rate at which technology is diffused worldwide depends heavily on the resource costs of transfer—both transmittal and absorption costs—and on the magnitude of the economic rents obtained by the seller. The resource costs of transfer depend on the characteristics of the transmitter, the receiver, the technology being transferred, and the institutional mode chosen for transfer.[2] These are matters explored in the following section.

1. S. Kuznets, *Modern Economic Growth: Rate, Structure, Spread.* (New Haven: Yale University Press, 1966).
2. The concept and measurement of the resource cost of transfer can be found in David Teece, *The Multinational Corporation and the Resource Cost of International Technology Transfer* (Cambridge: Ballinger, 1976), and in "Technology Transfer by Multinational Firms: The Resource Cost of International Technology Transfer," *Economic Journal* (June 1977).

The rents obtained are a function of the working of the market for know-how, a matter explored in a subsequent section. The last two sections explore regulatory issues with respect to this market from the perspective first of less-developed country (LDC) importers and from the perspective of the United States as a net exporter of know-how. What emerges is an understanding of the technology transfer process, the role of the multinationals, and the difficulties and occasional contradictions associated with regulation. In no sense can the market for know-how and the transfer process be said to operate in an ideal fashion. However, internalization of the process appears to offer considerable efficiencies, and "codes of conduct" are likely to confound the very objectives of importers, while export controls can be expected to yield only limited benefits, and then only under special conditions.

CODIFICATION AND TRANSFER COSTS

The fact that different individuals, organizations, or nations possess different types of knowledge and experience creates opportunities for communication and mutually profitable transfer. Yet, paradoxically, such transfer as does take place among individuals and organizations can only do so on the basis of similarities in the knowledge and experience each possess. A shared context appears necessary for the formulation of meaningful messages. Transmittal and receiving costs are lower the greater the similarities in the experience of the transmitting unit and the receiving unit; for the greater these similarities, the easier it is to transfer technology in codified form, such as blueprints, formulas, or computer languages.

Furthermore, there appears to be a simple but powerful relationship between codification[3] of knowledge and the costs of its transfer. Simply stated, the more a given item of knowledge or experience has been codified, the more economically it can be transferred. This is a purely technical property that depends on the ready availability of channels of communication suitable for the transmission of well-codified information—for example, printing, radio, telegraph, and data networks. Whether information so transferred will be considered meaningful by those who receive it will depend on whether they are familiar with the code selected as well as the different contexts in which it is used.[4]

Uncodified or tacit knowledge, on the other hand, is slow and costly to transmit. Ambiguities abound and can be overcome only when communications take place in face-to-face situations. Errors of interpretation can be corrected by a prompt use of personal feedback. Consider the apprenticeship system as an example. First, a master craftsman can cope with only a limited number of pupils at a time; second, his teaching has to be dispensed mostly through examples rather than by precept—he cannot easily put the intangible elements of his skill into words; third, the examples he offers will be initially confusing and ambiguous for his pupils so that learning has to take place through extensive and time-consuming repetition, and mastery will occur gradually on the basis of "feel"; finally, the pupil's eventual mastery of a craft or skill will remain idiosyncratic and will never be a carbon copy of his master's. It is the scope provided for the development of a personal style that defines a craft as something that goes beyond the routine and hence programmable application of a skill.

The transmission of codified knowledge, on the other hand, does not necessarily require face-to-face contact and can often be carried out largely by impersonal means, such as when one computer "talks" to another, or when a technical manual is passed from one individual to another. Messages are better structured and less ambiguous if they can be transferred in codified form. Take for example Paul Samuelson's introductory textbook for students of economics. Year after year, thousands of students all over the globe are introduced to Samuelson's economic thinking without being introduced to Samuelson himself. The knowledge acquired will be elementary and standardized, an idiosyncratic approach at this level being considered by many as a symptom of error rather than of style. Moreover

3. Codification—the transformation of experience and information into symbolic form—is an exercise in abstraction that often economizes on bounded rationality. Instead of having to respond to a hopelessly extensive and varied range of phenomena, the mind can respond instead to a much more restricted set of information. At least two obstacles stand in the way of effective codification. First, abstracting from experience can be accomplished in an almost infinite number of ways. Ask a group of painters to depict a given object and each will select different facets or features for emphasis. Furthermore, the choice of what to codify and how to codify it is often personal. Second, to structure and codify experience one way can make it difficult, subsequently, to do so in an alternative way. The conceptual channels through which experience is made to flow appear to run deep and resist rerouting.

4. These ideas are developed further in C. E. Shannon and W. Weaver, *The Mathematical Theory of Communication* (Chicago: University of Illinois Press, 1949). I am grateful to Max Boisot for drawing them to my attention.

the student can pick up the sage's book or put it down according to caprice; he can scan it, refer to it, reflect upon it, or forget it. This freedom to allocate one's attention or not to the message source is much more restricted where learning requires interpersonal contact.

With respect to the international transfer of technology, the costs of transfer are very much a function of the degree to which know-how can be codified and understood in that form by the recipient. Typically, only the broad outline of technical knowledge can be codified by nonpersonal means of intellectual communication or communication by teaching outside the production process itself. Accordingly, the transfer of technology generally requires the transfer of skilled personnel, even when the cultural and infrastructural differences are not great. History has illustrated this time and time again. For instance, the transfer of technological skills between the United States and Britain at the end of the nineteenth century was dependent upon the transfer of skilled personnel. One also observes that the diffusion of crafts from one country to another depends on the migration of groups of craftsmen, such as when the Huguenots were driven from France by the repeal of the Edict of Nantes under Louis XIV.

The costs of transfer so far examined are simply the resource costs of transfer—the costs of the labor and capital that must be employed to effect transfer. An empirical investigation of these issues based upon a sample of 26 international transfers indicated that the resource cost of international transfer is nontrivial.[5]

5. See David Teece, "Technology Transfer by Multinational Firms: The Resource Cost of International Technology Transfer," *Economic Journal* (June 1977).

Transfer costs ranged from 2.25 percent to 59 percent of total project costs with a mean of 19.16 percent. They declined with each subsequent application of the technology and were typically lower the greater the amount of related manufacturing experience possessed by the transferee. Experience with transfer and experience with the technology appear to be key considerations with respect to the ease with which technology can be transferred abroad. In order to understand these costs, as well as other aspects of the transfer process, it will be necessary to examine the market for know-how. In so doing, the focus is on private transactions between firms of different national origins.

CHARACTERISTICS OF THE MARKET
FOR KNOW-HOW

The differential distribution of know-how and expertise among the world's enterprises means that mutually advantageous opportunities for the trading of know-how commonly exist. However, these opportunities will be realized only if the institutional framework exists to provide the appropriate linkage mechanisms and governance structures to identify trading opportunities and to surround and protect the associated know-how transfers. Unfortunately, unassisted markets are seriously faulted as institutional devices for facilitating trading in many kinds of technological and managerial know-how.

The imperfections in the market for know-how for the most part can be traced to the nature of the commodity in question. Know-how has some of the characteristics of a public good, since it can often be used in another enterprise without its value being substantially impaired. Furthermore the marginal cost of employing know-how abroad is likely to be much less than its average cost of production and transfer. Accordingly the international transfer of proprietary know-how is likely to be profitable if organizational modes can be discovered to conduct and protect the transfer at low cost.

An examination of the properties of markets for know-how readily leads to the identification of several transactional difficulties. These difficulties can be summarized in terms of recognition, disclosure, and team organization. Consider a team that has accumulated know-how that can potentially find application in foreign markets. If there are firms abroad that can apply this know-how with profit, then according to traditional microeconomic theory, trading will ensue until the gains from trade are exhausted. Or, as Calabresi has put it, "if one assumes rationality, no transactions costs, and no legal impediments to bargaining, all misallocations of resources would be fully cured in the market by bargains."[6] However, one generally cannot expect this happy result in the market for proprietary know-how. Not only are there high costs associated with obtaining the requisite information, but there are also organizational and strategic impediments associated with using the market to effect transfer.

Consider the information requirements associated with using markets. In order to carry out a market transaction, it is necessary to discover potential trading partners and acceptable terms of trade. It is also necessary to conduct negotiations leading up to the bargain, to draw up the contract, to

undertake the inspection needed to make sure that the terms of the contract are being observed, and so on. As Kirzner has explained,

for an exchange transaction to be completed it is not sufficient merely that the conditions for exchange which prospectively will be mutually beneficial be present; it is necessary also that each participant be aware of his opportunity to gain through the exchange. . . . It is usually assumed . . . that where such scope is present, exchange will in fact occur. . . . In fact, of course, exchange may fail to occur because knowledge is imperfect, in spite of the presence of the conditions for mutually profitable exchange.[7]

The transactional difficulties identified by Kirzner are especially compelling when the commodity in question is proprietary information. One reason is that protecting the ownership of technological know-how often requires the suppression of information on exchange possibilities. By its very nature, industrial R&D requires that the activities and outcomes of the R&D establishment be disguised or concealed.

Even where the possessor of the technology recognizes the opportunity and has the capability to absorb know-how, markets may break down. This is because of the problems of disclosing value to buyers in a way that is convincing and that does not destroy the basis for exchange. Due to informational asymmetries, the less informed party must be wary of opportunistic representations by the seller. Moreover, if there is sufficient disclosure to assure the buyer that the information possesses great value, the "fundamental paradox" of information arises: "its value for the purchases is not known until he has the

6. G. Calabresi, "Transactions Costs, Resource Allocation, and Liability Rules: A Comment," *Journal of Law and Economics,* (April 1968).

7. I. Kirzner, *Competition and Entrepreneurship* (Chicago: University of Chicago Press, 1962), p. 215.

information, but then he has in effect acquired it without cost."[8]

Appropriability issues emerge not only at the negotiating state but also at all subsequent stages of the transfer. Indeed, as discussed elsewhere in this issue, Magee has built a theory of multinational enterprise around the issue of appropriability, hypothesizing that

multinational corporations are specialists in the production of information that is less efficient to transmit through markets than within firms.[9]

However, the transactional difficulties in the market for know-how that provide an incentive for firms to internalize technology transfer go beyond issues of recognition and appropriability. Thus suppose that recognition is no problem, that buyers concede value and are prepared to pay for information in the seller's possession, and that enforceable use restrictions soften subsequent appropriability problems. Even if these assumptions are satisfied, there is still the problem of actually transferring the technology.

In some cases the transfer of a formula or a chemical compound, the blueprints for a special device, or a special mathematical algorithm may be all that is needed to effect the transfer. However, more is frequently needed. As mentioned earlier, know-how cannot always be codified, since it often has an important tacit dimension. Individuals may know more than they are able to articulate.[10] When knowledge has a

high tacit component, it is extremely difficult to transfer without intimate personal contact, demonstration, and involvement. Indeed, in the absence of intimate human contact, technology transfer is sometimes impossible. In a slightly different context Polanyi has observed, "It is pathetic to watch the endless efforts—equipped with microscopy and chemistry, with mathematics and electronics—to reproduce a single violin of the kind the half literate Stradivarius turned out as a matter of routine more than 200 years ago."[11]

In short, the transfer of knowledge may be impossible in the absence of the transfer of people. Furthermore, it will often not suffice just to transfer individuals. While a single individual may sometimes hold the key to much organizational knowledge, team support is often needed, since the organization's total capabilities must be brought to bear upon the transfer problem. In some instances the transfer can be effected through a one-time contract providing for a consulting team to assist in the start-up. Such contracts may be highly incomplete and may give rise to dissatisfaction during execution. This dissatisfaction may be an unavoidable—which is to say, an irremediable—result. Plainly, foreign investment would be a costly response to the need for a one-time international exchange. In the absence of a superior organizational alternative, one-time, incomplete contracting for a consulting team is likely to prevail.

Reliance on repeated contracting is less clearly warranted, however, where a succession of transfers is contemplated, or when two-way communication is needed to promote the recognition and disclosure of opportunities for information

8. K. J. Arrow, *Essays in the Theory of Risk Bearing* (Chicago: Chicago University Press, 1971).

9. See Stephen Magee, "Information and Multinational Corporation: An Appropriability Theory of Direct Foreign Investment," in *The New International Economic Order*, ed. Jagdish Bhagwati (Cambridge, MA: MIT Press, 1977), p. 318.

10. See Michael Polanyi, *Personal Knowledge: Towards a Post Critical Philosophy* (Chicago: University of Chicago Press, 1958).

11. Polanyi.

transfer as well as the actual transfer itself. In these circumstances a more cooperative arrangement for joining the parties would enjoy a greater comparative institutional advantage. Specifically, intrafirm transfer to a foreign subsidiary, which avoids the need for repeated negotiations and attenuates the hazards of opportunism, has advantages over autonomous trading. Better disclosure, easier agreement, better governance, and more effective team organization and reconfiguration all result. Here lies the incentive for internalizing technology transfer within the multinational firm.

The preceding discussion has emphasized that an important attribute of the multinational firm is that it is an organizational mode capable of internally transferring know-how among its various business units in a relatively efficient and effective fashion. Given the opportunities that apparently exist for international trade in know-how, and given the transactional difficulties associated with relying on markets, one should expect to find multinational enterprises (MNEs) frequently selecting internal channels for technology transfer. However, when problems of recognition, disclosure, and team transfer are not severe, one should expect that market processes will be utilized, in which case the licensing of know-how among nonaffiliated enterprises will be observed.

Recognition, disclosure, and team transfer problems will be modest, it would seem, when the following exist: (1) the know-how at issue is not recent in origin so that knowledge of its existence has diffused widely; (2) the know-how at issue has been commercialized several times so that its important parameters and performance in different situations are well understood, thereby

reducing the need for start-up assistance; and (3) the receiving enterprise has a high level of technological sophistication. Some evidence supportive of these propositions has recently been presented. Mansfield, employing a sample of 23 multinationals, discovered that foreign subsidiaries were the principal channel of transfer during the first five years after commercialization.[12] For the second five-year period after commercialization, licensing turned out to be more important. Larger firms also tended to rely more on internal transfer than did smaller firms, although this might not reflect relative efficiency considerations but rather the sunk costs larger firms have already made in foreign subsidiaries.

One implication for a potential technology purchaser is that it will have to look among the smaller firms in the industry, and at firms in different industries, to find willing technology suppliers. This does not result in an easy search process. It is made more difficult by the fact that few firms actively market their know-how. Thus the apparent size and nature of the market is likely to be a function of the search costs buyers are willing to incur.

Another implication is that because the marginal cost of successive applications of a technology is less than the average cost of production and transfer, and because know-how is often unique—implying that trading relations are

12. See Edwin Mansfield, "Statement to the Senate Commerce Committee Concerning International Technology Transfer and Overseas Research and Development," Hearings before the Subcommittee on International Finance of the Committee on Banking, Housing, and Urban Affairs of the Committee on Commerce, Science, and Transportation, United States Senate, Ninety-fifth Congress, Second Session, Part 7: Oversight on U.S. High Technology Exports (Washington, DC: Government Printing Office, May 1978).

characterized by small numbers—there is often a high degree of indeterminacy with respect to price. Killing's field research confirmed that "neither buyer nor seller of technology seems to have a clear idea of the value of the commodity in which they are trading," fueling speculation that "royalty rates may simply be a function of negotiating skills of the parties involved."[13] This is because the market for know-how commonly displays aspects of bilateral monopoly, at least at the level of the individual transaction. So in many important cases there is likely to be a wide range of indeterminacy.

The existence of elements of bilateral monopoly has led some countries to advocate regulation of the market for know-how. Indeed, some Third World countries, as well as the antitrust authorities in some developed countries, have already imposed various regulatory regimes.

By 1974, over 20 countries had enacted specific legislation to control and direct foreign capital and technology. Their actions and regulations focused on lowering the royalties paid for foreign technology, forcing local participation in management and ownership, and in increasing the government capability to screen and direct foreign activities—the major focus of the governments was initially to limit the kind of restrictive clauses allowed in contracts for technology transfer with foreign firms.[14]

Governmental and intergovernmental intervention in the market for

13. Peter Killing, "Technology Acquisition: License Agreement or Joint Venture," *Columbia Journal of World Business* (Fall 1980).

14. See Harvey Wallender, "Developing Country Orientations Towards Foreign Technology in the Eighties: Implications for New Negotiation Approaches," *Columbia Journal of World Business* (summer 1980): 21-22.

know-how appears to be growing in significance. In the following sections, several dimensions of this phenomenon are explored in more detail.

CODES OF CONDUCT AND THE REGULATION OF TECHNOLOGY IMPORTS

Since the United Nations Conference on Trade and Development (UNCTAD) IV decided to set up an intergovernmental group of experts to prepare a draft of an international code of conduct on the transfer of technology, discussion has intensified on matters associated with the transfer and development of technology, particularly on topics of concern of developing countries.[15] A number of draft codes have emerged in which representatives from less-developed countries have argued that technology is part of the universal heritage of mankind and that all countries have right of access to technology in order to improve the standards of living of their peoples. Such contentions obviously involve fundamental challenges to the world's industrial property system. They also fail to recognize the constitutional restraints in countries such as the United States that pre-

15. The movement toward an international code on the transfer of technology is but a reflection of larger, exceedingly complex political problems that have been engendered by an international society undergoing profound changes. Demands for a new international economic order, international regulation of transnational enterprises, and the like form the backdrop of UNCTAD's activities in the technology transfer area. These broader demands raise the possibility that the work now being carried on by UNCTAD in moving toward a code of conduct for the transfer of technology will be subsumed by the development of a more comprehensive code of conduct for transnational enterprises by the U.N. Commission on Transnational Corporations.

vent the government from confiscating private property.

The stated objective of the UNCTAD code is "to encourage the transfer of technology transactions, particularly those involving developing countries, under conditions where bargaining positions of the parties to the transaction are balanced in such a way so to avoid abuses of a stronger position and thereby to achieve mutually satisfactory agreement." One of the principal mechanisms by which this is to be achieved is through the elimination of "restrictive business practices."[16] A long litany of these is typically advanced, including tying or packaging, use restrictions, exclusive dealing, and territorial restrictions. An examination of recent legislation on the transfer of technology, particularly in Latin America and Yugoslavia, shows that many of these ideas have been uncritically accepted into national law.[17]

It is not possible to attempt a comprehensive review of restrictive business practices in this article. However, I submit that insufficient analysis has been given to the efficiency-enhancing attributes of many practices surrounding the generation and transfer of technology. Many restrictive clauses in licensing and know-how agreements are designed to protect the transaction and the underlying know-how; in their absence less technology might be transferred, to the mutual detriment of all, or technology might be transferred less efficiently. In the space that follows,

two "restrictive business practices" —use restrictions and tying—are analyzed in order to illustrate that "restrictive business practices" can be in fact procompetitive and may serve to promote economic efficiency.

Use restrictions

The interesting question associated with use restrictions is whether they are anticompetitive, designed merely to extract monopoly rents, or whether they are efficiency instruments, the removal of which might leave both parties worse off. Since know-how is the principal resource upon which the value of many private enterprise firms is based, firms facing market competition are not going to sell it carte blanche to a firm that might use it to compete with their own products, for to do so would reduce the value of the firm. Thus reasonable limitations on use are commonly necessary to provide adequate incentives for transfers to occur and for those transfers to operate efficiently. This is especially true when the transferor and the transferee are competitors or potential competitors.

When know-how is transferred by a market transaction (contract) the buyer does not acquire the asset to the exclusion of use by the seller in the same sense as occurs when a physical item is bought and sold. The seller of know-how retains the knowledge even after it has been transferred to a buyer. Furthermore, technology is constantly evolving. Indeed, static technology is generally obsolete technology. Accordingly, a buyer of intangible know-how typically needs ongoing, future cooperation from the seller to obtain the full benefit of the know-how purchased, since all of the

16. See UNCTAD, "Draft International Code of Conduct on the Transfer of Technology," TD/CODE/TOT/20.

17. See UNCTAD, "Selected Legislation, Policies and Practices on the Transfer of Technology," TD/B/C.6/48.

learning and experience of the developer of the know-how cannot be captured in the codified descriptions, drawings, and data that are amenable to physical transfer.

Limitations on the use of technological know-how are often needed to provide adequate incentives for the buyer and the seller to effect a continuous transfer of the knowledge in question. If the seller is limited in his use of the know-how, the buyer can rely more confidently on the seller's full disclosure and cooperation in the buyer's use of the know-how. Where the seller contemplates some use of the know-how himself, limitations on the buyer's use of the know-how in competition with the seller are necessary to provide the seller with the incentive to transfer this know-how and to share fully in his mental perceptions, understandings, working experience, and expertise.

A partial analogue to these principles is when business enterprises are sold. These transactions have traditionally included ancillary limitations on the economic activities of the seller after the business is sold. Such limitations bring about economically efficient transfers of ongoing businesses by ensuring that the buyer acquires exclusively the enterprise—or part of the enterprise —he is contracting to purchase, including its intangible goodwill. Similarly, in the sale of a business the seller is often retained as a consultant for the purpose of ensuring that the intangible knowledge that comes from the seller's experiences in conducting the business is fully transferred in the transaction. Without contractual or other limitations on the seller's use of the assets being transferred, and without the seller's continued cooperation, a buyer would not pay the full eco-

nomic value of those assets. As a result, the efficient transfer of the assets would be inhibited.

Use limitations are particularly beneficial when two or more uses exist for the products that can be derived from know-how and when some of the uses are for some reason foreclosed to the developer of the know-how. In this instance, transfer of the know-how to a buyer having access to one or more of these otherwise foreclosed uses may be beneficial to both parties, since economies of scope will be generated. The seller of the know-how requires adequate incentives to transfer his knowledge, however. The seller will not transfer the know-how to a buyer for the otherwise foreclosed uses if, in doing so, he is likely to lose more in the uses that are available to him with no transfer than he gains through the expanded uses made possible by transfers. The availability of limitations on the buyer's use of the know-how provides possible means to prevent such losses.

Use limitations are also beneficial in providing incentives for the contracting parties to share complementary know-how in order to reach a new market that neither acting independently could efficiently serve. If each of the parties has one or more of the technology elements critical for a particular new use, if neither of the parties has all of the critical technology elements for that use, and if through sharing of the complementary technologies for the new use one or both of the parties could enter markets that neither party could serve without sharing, then use limitations are necessary to effect the bilateral technology transfers. Without use limitations, one or both of the parties may lack the incentive to share, since the losses that might occur in an existing market through sharing could exceed the gains derived from reaching the new market.

Tying and packaging

In a tying arrangement, the seller requires the buyer to purchase a second product as a condition of sale of the first, such as when a petrochemical firm licenses its process technology to another firm on the condition that it purchase certain inputs on a continuous basis, or when an automobile company agrees to build a facility abroad so long as it is able to select equipment and designs for the whole facility and not just for part of it.

In the context of the international transfer of technology, there are often very genuine managerial and technical reasons for tying the sale of products. For instance, coordinated design and construction might allow important systems engineering functions to be carried out more efficiently. Furthermore, processing facilities may require raw materials and components that meet certain narrow technological standards, and tying may be necessary to ensure that the requisite amount of quality control is exercised. These problems are likely to be especially severe when the technological distance between the transferor and transferee is great.

It is only under rather special circumstances that tying will enable a monopolist to expand the amount of monopoly profit that would be obtained in the absence of tying. One such circumstance is if tying can be used as a method of price discrimination.

Accordingly, blanket prohibitions against tying and packaging are likely to be costly to the country imposing the prohibitions. Technol-ogy suppliers may have good reasons for wanting to supply know-how and other products and services in a package. Certainly some striking examples exist of problems that have arisen when adequate packaging and systems design have not been performed. Consider the Soviet Union's experience in constructing and starting up its Kama River truck plant, as related by Lee Iacocca, then with Ford Motor Company:

Well, one example of acquiring technology in its unbundled state is the Kama River truck plant in Russia. After first attempting to get a foreign company to build the plant (we were approached but decided against it) the Russians decided to do it themselves and to parcel out contracts to foreign firms for various parts of the project. That was in 1971. As of December 1976, the project was almost two years behind schedule. By year's end, only about 5,000 trucks were expected to roll off the line, instead of the 150,000 vehicles and 100,000 diesel engines and transmissions originally scheduled for annual production. According to published reports, only four of nine projected furnaces in the iron foundry were operating and those only at half capacity. What's more, 35 percent of the castings were being rejected as unserviceable. There were bottlenecks on the assembly line, and because the components and designs were bought from different suppliers all over the world, replacement parts were not interchangeable.

Now compare that with Ford's recent investment in Spain. It took us just three years to the day to build a complex that includes an assembly plant, a stamping and body plant and an engine plant on a manufacturing site 2 1/2 miles long and half a mile wide, with 55 acres under roof. The first Fiesta, our new minicar, was driven off the assembly line last August, well ahead of schedule. To get from farmland to an annual capacity of 250,000 cars and 400,000 engines in three years, we drew on the experience of our personnel and our technological resources from all over the world—experience and resources that couldn't be bought and that we probably wouldn't even know how to sell.[18]

REGULATION OF TECHNOLOGY EXPORTS

Pressures for restricting trade in technological know-how have also come from technology exporters. The reasons advanced for controls are almost the complete opposite of those advanced by the LDCs. In the United States concern is often expressed in industry and government that the United States is either selling its technology for far less than its economic value, or allowing it to be stolen through industrial espionage, principally to other developed countries, or simply transferring it abroad too soon. According to J. Fred Bucy, the president of Texas Instruments:

Today our toughest competition is coming from foreign companies whose ability to compete with us rests in part on their acquisitions of U.S. technology. . . . The time has come to stop selling our latest technologies, which are the most valuable things we've got.[19]

Labor groups in the United States go further and argue that not only is the know-how underpriced, but that one consequence of the export of technology is the export of jobs.[20] According to one labor leader:

I recognize that technology will flow across national lines no matter what we

18. See Lee Iacocca, "Multinational Investment and Global Purpose," speech delivered before the Swiss-American Chamber of Commerce," Zurich, June 17, 1977. Reprinted in *Vital Speeches*, 15 Sept. 1977.
19. See "Those Worrisome Technology Exports," *Fortune*, 22 May 1978, p. 106.
20. An example commonly cited is that of Piper aircraft. Until a few years ago, Brazil

do. But certainly we do not have to cut our own throats with aid, trade, tax and tariff policies that actively encourage and promote the export of American jobs and technology, without regard for the impact on either those who give or those who receive.[21]

Before proceeding further, it will be helpful to outline the available evidence with respect to these considerations. Unfortunately, only very sketchy data are available. Conclusive evidence on the net impacts of foreign investment and technology transfer on U.S. jobs and welfare does not exist. The available evidence suggests that the impact is likely to vary from one instance to another. Baranson has presented case studies that suggest that U.S.-based firms, driven by competitive necessity, are transferring their newest technology abroad more frequently than in the past.[22] To investigate this issue further, Mansfield and Romeo obtained information concerning the age of the technology transferred abroad in a sample of 65 transfers taken from 31 U.S.-based

was the leading purchaser of light aircraft manufactured in the United States. However, the Brazilian government levied prohibitive taxes on the import of American-produced light aircraft and it invited an American manufacturer, Piper, to bring in U.S. technology and produce with Brazilian workers. As a result, hundreds of U.S. citizens who were directly employed in light aircraft production became unemployed, some permanently. Now Brazil is selling light aircraft to other Latin American countries and is also planning to export planes to the United States in competition with American producers.
21. William Winpisinger, "The Case Against Exporting U.S. Technology," *Research Management* (March 1978): 21.
22. Jack Baranson, "Technology Exports Can Hurt," *Foreign Policy*, 25 (Winter 1976-77).

TABLE 1
MEAN AND STANDARD DEVIATION OF NUMBER OF YEARS BETWEEN
TECHNOLOGY'S TRANSFER OVERSEAS AND ITS INITIAL INTRODUCTION IN
THE UNITED STATES, FOR 65 TECHNOLOGIES

CHANNEL OF TECHNOLOGY TRANSFER	MEAN (YEARS)	STANDARD DEVIATION (YEARS)	NUMBER OF CASES
Overseas subsidiary in developed country	5.8	5.5	27
Overseas subsidiary in developing country	9.8	8.4	12
Licensing or joint venture	13.1	13.4	26

SOURCE: Edwin Mansfield and Anthony Romeo, "Technology Transfer to Overseas Subsidiaries by U.S.-Based Firms," Research Paper, University of Pennsylvania, 1979.

multinationals.[23] As shown in Table 1, they found that the mean age of the technologies transferred to overseas subsidiaries in developed countries was about 6 years, which was significantly less than the mean age of technologies transferred to overseas subsidiaries in developing countries—about 10 years. Table 1 also suggests that the mean age of the technologies transferred through licenses, joint ventures, and channels other than subsidiaries is commonly higher than the mean age of the technologies transferred to subsidiaries, indicating that firms tend to transfer their newest technology overseas through wholly owned subsidiaries rather than via licenses or joint venture, but the latter channels become more important as the technology becomes older.

Another concern of countries that generate new technology is that the transfer of technology to overseas subsidiaries will hasten the time when foreign producers have access to this technology. Some evidence has recently become available on the speed with which technology "leaks out" and the extent to which international transfer actually hastens its "leaking out." The evidence, which is based on a sample of 26 technologies transferred abroad, indicated that the mean lag between the transfer and the time when foreign firms had access to the technology was about four years.[24] In over half the cases, the technology transfer was estimated to have had no effect at all on how quickly foreign competitors had access to the technology. On the other hand, in about one-fourth of the cases, it was estimated to have hastened their access to the technology by at least three years.

Technology transfer hastened the spread of process technologies to a greater degree than it did the spread of product technologies. According to the study, the most frequent channel by which the technology "leaked out" was reverse engineering.[25] That is, foreign com-

23. See Edwin Mansfield and Anthony Romeo, "Technology Transfer to Overseas Subsidiaries by U.S.-Based Firms," Research Paper, University of Pennsylvania, 1979.

24. Ibid.
25. Reverse engineering is very common in the semiconductor industry. it involves

petitors took apart and analyzed the new or modified product to gain insights into the relevant technology. Clearly, this evidence gives only a very sketchy impression of the level and nature of the returns from international technology transfer, and the role that technology exports are having on the U.S. competition position. However, there is little evidence that the technological lead of the United States in various industries is about to disappear as a result of the technology transfer activities of American firms. Indeed, there is some evidence, admittedly of a conjectural nature, that the international transfer of technology stimulated R&D activities by multinational firms.[26]

From a public policy perspective the interesting question is whether the United States could increase its economic welfare through restrictions on technology exports. It is a well-known theorem of interna-

tional trade that if a country has monopoly (monopsony) power in world markets, then imposing a tax (tariff) on exports (imports) will serve to improve welfare in the absence of retaliation. This, of course, assumes that such a policy can be effectively administered.

The economic intuition behind this theorem is fairly apparent. By transferring technology abroad, American firms increase the likelihood of foreign competition in the future. While firms face incentives to consider this when setting prices at which technology is transferred, each firm will evaluate the future effects on themselves, not on the rest of the economy. The company that exports the technology is not usually the one that loses out. It receives payment of some kind. The victim is likely to be another American company, one that prior to the technology transfer enjoyed a competitive advantage over the foreign company. Fujitsu, for example, has used the technology it got from Amdahl to compete with IBM. Therefore, in strictly nationalist terms, private firms will have a tendency to set the price of technology too low and to transfer too much technology abroad. Where several U.S. firms have similar technology that does not exist abroad, their competition will tend to lower the price of technology transfers. The United States could prevent this by reducing competition and by establishing monopoly prices through control of such transfers. For instance, an export tax would serve to restrict exports, thereby driving up the price and enabling the United States economy to capture monopoly rents from the export of know-how. A similar result could be obtained by enabling

---

stripping down a competitor's chip to recreate an outright copy, to figure out how a chip works in order to design a functionally equivalent emulator chip, or merely to determine whether a new chip contains any new ideas that might be adaptable to other products. Creating a copy is surprisingly simple: the necessary tools include a microscope, acid to etch away the circuits layer by layer, and a camera to record the successive steps; $50,000 of equipment will suffice. Reverse engineering enables a rival to obtain the same advantages as could be obtained by pirating the masks—the negatives that are used to lay down the circuit elements on silicon wafers—used in manufacturing the product. Intel Corporation of California has accused the Soviet Union of copying one of its 4K memory chips and Japan's Toshiba Corporation of making a "dead ringer" of another. See *Business Week*, 21 April 1980, p. 182.

26. See Edwin Mansfield, Anthony Romeo, and Samuel Wagner, "Foreign Trade and U.S. Research and Development," *Review of Economics and Statistics*, 1979.

domestic industry to cartelize foreign markets.[27]

There is, in fact, a long history of government attempts to limit the export of technology and trade secrets. A prime example is England during the Industrial Revolution. There are serious disadvantages in limiting technology transfers, however. One problem is that while levels of restriction that are optimal on nationalist grounds can be determined in theoretical models, there is little reason to be confident that government policies will approach such optima in practice. Domestic firms seem able to circumvent restrictions on the export of know-how,[28] while foreign firms can engage in "reverse engineering of products and designs" to circumvent many controls.

An alternative approach to technology controls might involve placing more emphasis on technical data and critical manufacturing equipment and less on commodities.[29] However, it is enormously difficult to control the export of technical data, since it can move in many informal ways that are often difficult to detect.[30] Clearly the transfer of highly visible turnkey plants is more readily controlled than are surreptitious, casual conversations. Furthermore, the effectiveness of controls depends on the degree of monopoly power possessed by the United States. In most instances where controls are applicable, the United States does not have a clear superiority vis-a-vis other Western countries. The effectiveness of controls therefore depends upon cooperation with other suppliers and potential suppliers.[31]

CONCLUSION

In the foregoing discussion, the arms-length market for know-how has been shown to be exposed to a number of hazards and inefficiencies, many of which can be overcome by internalizing the process within the multinational firm. Despite the shortcomings identified, it was not apparent that regulation by either technology importers or exporters could substantially improve the efficiency with which this market operates; indeed, for the instances examined it appeared that the impairment of efficiency through regulation was the more likely outcome. Yet the strongest argument against controls on the transfer of

27. One difference is that with a cartel as compared with a tax, the industry would capture a larger portion of the economic rents, as there would be no revenues accruing to the government.

28. In 1980, allegations of export control violations in the United States numbered 350, up from 200 in 1979. *Business Week*, 27 April 1981, p. 131.

29. On the other hand, some authorities suggest that Soviet spies might do better acquiring consumer products in large department stores. Buying consumer and industrial products such as toys, appliances, and industrial tools in many cases may be more useful than technical data because of the delays in Defense Department procurement of new chips and the rapidity with which new chips become incorporated into consumer products.

30. According to one source, the KGB has 30 agents in California's Silicon Valley, plus others in Phoenix and Dallas, charged with obtaining data on microeconomics technology. *Business Week*, 27 April 1980, p. 128.

31. The Coordinating Committee on Export Controls (COCOM), an organization consisting of all NATO members plus Iceland and Japan, is the forum usually chosen to attempt the necessary coordination. However, the members have no legal obligation to participate in COCOM or to abide by its recommendations.

technology is the same as the argument for liberal trade policies in general. Many kinds of economic restrictions can be used to bring gains to some at the expense of others. But almost everyone is likely to end up worse off if they all succeed. This holds just as true for nations within the world economy as for individuals and groups within a national economy. The basic case for liberal policies is not that they always maximize short-run gains, but that they serve enlightened and longer-run interests in avoiding a world riddled with restrictions.

THE AMERICAN ACADEMY OF POLITICAL AND SOCIAL SCIENCE
– ANNUAL MEETING 1982 –

TOPIC: INTERNATIONAL TERRORISM

CALL FOR PAPERS

As part of The Academy's continuing effort to promote dialogue among its membership, the 1982 Annual Meeting will, for the first time, be open to a select number of papers to be chosen from abstracts submitted by Academy members. This meeting, to be held in Philadelphia on April 23–24, 1982, will center on the topic of international terrorism. Papers presented are to be 2500 to 4000 words in length. Interested members should submit an abstract (not to exceed 200 words) and a one-page *curriculum vitae* to:

> Marvin E. Wolfgang
> President
> The American Academy of Political and Social Science
> 3937 Chestnut Street
> Philadelphia, PA 19104

Deadline for abstracts: January 15, 1982

Deadline for submission of chosen papers: March 1, 1982

The Academy will provide local hospitality during the meeting for those whose papers are selected.

Arguments for the Generation of Technology by Less-Developed Countries

By FRANCES STEWART

ABSTRACT: This article considers whether LDCs should try to develop their own technological capacity rather than continue to rely on developed countries for their technology. The developed countries at present have a comparative advantage in the production of technology. They have the specialized resources of manpower and physical equipment as well as considerable experience. It is therefore likely to be cheaper for developing countries to buy their technology from developed countries. But there are three reasons why developing countries should nonetheless try to develop their own technological capacity. First, local technological capacity is necessary to adapt imported technology to local conditions so that it becomes more efficient in use. Second, technology imported from industrialized countries is often inappropriate for the different conditions in less-developed countries (LDCs). The imported technology tends to be capital-intensive and large scale and often produces oversophisticated high income products. LDCs need their own technological capacity to develop more appropriate technologies. Third, total technological dependence is a major factor behind the generally dependent relationship that many countries experience vis-a-vis industrial countries. Without independent technological capacity, the technological dependence and the more general dependent relationship is likely to be perpetuated.

Frances Stewart is a fellow of Somerville College and senior research officer at the Institute of Commonwealth Studies, Oxford. She is the author of Technology and Underdevelopment *(Macmillan, 1977) as well as of many articles on questions of technology, employment, and the international monetary system. She has been a consultant to many organizations, including the World Bank, the International Labor Organization, the United Nations Conference on Trade and Development, and the United Nations Industrial Development Organization. She is a director of the Intermediary Technology Development Group.*

THE generation of technology in the modern sector is dominated by the developed countries. The main purpose of this article is to consider the arguments that would justify less-developed countries (LDCs) trying to create their own technological capacity; in doing so we also come to some tentative conclusions about the kind of technological capacity that LDCs might aim to promote. First we consider, as a background, the nature of developed country (DC) domination of technological development.

It is helpful to classify technology in a threefold division: products, machines/processes to make the products we describe as the "hardware production technology," and software services necessary for the production/consumption of these products. Thus in any particular productive activity we have a particular product—for example, a car—production techniques to manufacture the car and various ancillary services, such as insurance, accountancy, management and so on, associated with production and consumption of the final product.

In the traditional sector, technology—in all three aspects—has been created over a very long period in and for the locality in which it operates. In the modern sector, where technological change is of much greater importance, technological development is largely dominated by the developed countries.

The generation of technology is a complex process that may result from formal research and development or may be the result of more informal "home" inventors and shopfloor adaptations and developments of existing techniques. Historically, informal sources of technological development were dominant[1]; the institution of research and development departments and the systematic application of science to the generation of technology occurred only in the last quarter of the nineteenth century. Since then there has been a dramatic growth of research and development (R&D) in developed countries. Nonetheless small improvements in technology, often the result of informal shopfloor activities, remain significant, accounting for perhaps one-half of total innovations.[2]

During the past 200 years technological innovations have been dominated by a handful of developed countries. A study conducted by the Organization for Economic Cooperation and Development (OECD) in 1970 identified 112 significant innovations in the twentieth century: all emanated from developed countries, with the United States responsible for 60 percent; the United Kingdom, 14 percent; and German firms, 11 percent.[3] These figures reflect past dominance by a handful of developed countries, but much of this domination remains. The developed countries are estimated to account for 97 percent of world expenditure on R&D, although this must be qualified to allow for the lower salaries

1. See D. C. Landes, *The Unbound Prometheus*, (Cambridge: Cambridge University Press, 1969).

2. See the evidence collected about innovation in Du Pont by S. Hollander, *The Sources of Increased Efficiency, A Study of the Du Pont Rayon Plant* (Boston: MIT, 1966); the research directed by J. Katz (J. Katz et al., "Productivity, Technology and Domestic Effects in Research and Development," IBD/ECLA Working Paper, No. 13); and the evidence of J. Enos, "Invention and Innovation in the Petroleum Refining Industry," in *The Rate and Direction of Inventive Activity* (Princeton: Princeton University Press, 1962).

3. OECD, 1970.

paid scientific manpower in LDCs; developing countries account for 13 percent of world scientists and engineers engaged in research and development.[4] Of the three and a half million patents issued in 1972, only six percent were issued by developing countries.[5]

In recent years there has been some increase in technological capacity among developing countries, an increase that is unevenly spread among countries. This is indicated by rising expenditure on R&D and by some evidence of technology exports by a few countries.[6] But while these developments are potentially significant, the general domination of technological development by developed countries remains. Among five countries with science and technology plans, four emphasized the weak links between research and economic activity.[7] According to Ranis, "the less developed world is strewn with scientific institutes and other expensive white elephants which contribute neither to science nor to technology."[8]

While the DC domination of formal R&D is well established, there is a quite substantial amount of LDC informal technology generation. Its extent varies substantially among countries, and we have very little systematic research to establish its quantitative significance. Nonetheless detailed firm studies,[9] as well as the evidence from South-South technology transfer,[10] suggest that in some cases this form of technology adaptation may be of quite substantial importance.

The threefold classification of technology becomes relevant here. For the most part, formal R&D is especially related to product innovation, and while process/hardware innovation often—perhaps nearly always—follows, product innovation provides the dominant motivation.[11] Informal technology generation is of greatest significance in terms of adapting technology hardware to improve its efficiency in a particular locality. Historically, software technology has rarely been a lead sector—in terms of technology change—rather, it is designed and introduced to suit the product/hardware technology in use. This is changing rather dramatically with the microelectronic revolution, whereby the radically changed possibilities in software—management, control, and communications—permit and indeed are leading to changes in products and in hardware. This revolution is largely of formal origin.

4. J. Annerstadt, "Technological Dependence: A Permanent Phenomenon of World Inequality," mimeo, Institute of Social and Economic Planning, Roskilde University Center, 1978).

5. UNCTAD, "The Role of the Patent System in the Transfer of Technology to Developing Countries," TD/B/AC.11/19, Rev. 1, 1975.

6. See evidence collected by S. Lall, *Developing Countries as Exporters of Technology* (London: Macmillan, forthcoming).

7. UNCTAD, "Technology Planning in Developing Countries: A Preliminary Review," TD/B/C.6 29, 1978.

8. In W. Beranek and G. Ranis, *Science and Technology and Economic Development* (New York: Praeger, 1978). See also D. Crane, "Technological Innovation in Developing Countries: A Review of the Literature," *Research Policy*, 6 (1977).

9. See for example P. Maxwell, "Learning and Technical Change in the Steelplant of Acinder S.A. in Rosario, Argentina," BID/CEPAL/BA/18, 1976.

10. Lall, *Developing Countries.*

11. See surveys on motivation for research and development; for example, Gustafson, "Research and Development, New Products and Productivity Change," *American Economic Review, Papers and Proceedings*, LII (1962).

It follows that although formal R&D may be responsible for only a proportion of the total innovation, it is the dominant source of innovation today, in the sense that it largely determines changes in products and software. Informal innovation tends to be adaptive rather than revolutionary and to work within the context set by formal R&D. Thus to the extent that DCs dominate in terms of effective formal R&D they tend to determine the general direction of technical change. Despite the existence in some LDCs of a considerable capacity for informal technology generation, the general situation remains therefore one of prime dominance by DCs.

Is this situation necessarily to the disadvantage of LDCs? It has long been argued that there are considerable advantages of being a "latecomer" in terms of industrial and technical development.[12] The latecomers may make use of the vast accumulation of scientific and technological knowledge for which the industrialized countries had to devote considerable resources. Yet the newly industrializing countries get this knowledge at relatively low cost—a good deal is freely available; for other parts a price is charged, but the price is much lower than that which would be necessary if LDCs were to develop the technology for themselves. But it does not follow from the "latecomer" argument that LDCs should remain dependent on the industrialized countries for technology, rather that they start in a relatively favorable position for developing technology.

Assuming that LDCs do wish to industrialize now—and not to postpone industrialization indefinitely to reap even more advantages from being a latecomer—they must choose whether to industrialize on the basis of technology "transferred" —or more accurately, bought—from industrialized countries or whether to generate their own technology, or some combination. One way of looking at this problem is from the point of view of comparative advantage; another viewpoint is that of cost-benefit analysis. Both would seem to justify continued technological dependence.

Comparative advantage—today—would seem to rest with the developed countries in the generation of technology. They have the major resources that make for efficient development of technology. The have the human resources of scientists and technologists, the laboratories, and the experience, all on a scale sufficient to exploit the economies of scale in technology production. Moreover in many modern products there are substantial indivisibilities in R&D, especially development, so that a large scale of production is necessary if R&D costs do not become exorbitant. Casual consideration is enough to suggest that LDCs' comparative advantage lays elsewhere. More detailed investigation of the effectiveness of R&D expenditure and other innovatory activity would on the whole support this view.[13]

Cost-benefit analysis of technology production in LDCs tends to lead to a similar conclusion. Comparing the costs of LDCs in producing a technology themselves with the costs of buying it from DCs generally shows big cost advantages in buying it from DCs, and this is true despite the fact that for many tech-

12. A. Gerschenkron, *Economic Backwardness in Historical Perspective* (Cambridge, MA: Harvard University Press, 1962).

13. See Beranek and Ranis, *passim*.

nologies, monopolistic elements in the market mean that the LDCs pay more than they would in more competitive conditions. The big differences in cost are not simply a reflection of differences in cost of producing technology in DCs and LDCs—arising from comparative advantage—but rather that LDCs do not pay for the costs of the production of the technology, since they are "marginal" purchasers, but for a sum covering communication/transfer costs and any monopolistic rents the sellers are able to levy. Since most of the technologies LDCs wish to use are not "frontier" technologies but technologies that have already been in use for some time in DCs, they rarely need to pay the production cost for the technologies acquired.

Another way of putting these arguments is that LDCs—by remaining somewhat behind the frontier countries—are able to continue to reap advantages of being latecomers, in terms of using older, already extant technology, even after they have started to industrialize. A good deal of Japanese technology policy has been based on this principle; even now, Japan relies heavily on technology bought from the West for its prime technology input and in so doing is able to get it cheaper than if it were to produce the technology from scratch.[14]

Despite the force of these arguments which would favor continued reliance on DC technology, the case for independent generation of technology is strongly felt among many

LDCs. Some powerful arguments can be adduced for this case. These arguments fall into three classes. First, some technological capacity may be necessary to make efficient use of imported technology. From this point of view it is not a question either of importing technology or of creating local capacity to generate technology, but rather that both are necessary for efficient production. Second, relying on Northern technology may mean relying on increasingly inappropriate technology: Southern generation of technology may be the only effective way of creating efficient appropriate technology. Third, there is the belief that technological dependence is a critical part of the generally dependent relationship between North and South; some degree of technological independence is then a necessary condition for achieving a more equal relationship. Although there are connections between the three sets of arguments, they are each rather different, with different implications for the type of technological capacity it is desirable to create. We shall consider the arguments one by one.

"TECHNOLOGICAL MASTERY"[15]

To transfer and make efficient use of imported technology undoubtedly requires considerable "technological mastery" or technological capacity. The technology has to be adapted to local conditions—to differences in the quality/availability of various factors, in government regulations, market conditions and

14. The first year Japan had a surplus on her external "technology account" with receipts from royalties and licenses exceeding payments was 1980. It is believed that this reflects continued large deficits with the developed countries which is counterbalanced by receipts from developing countries. See *Financial Times*, 12 June 1981 reporting on a survey in the *Japanese Economic Journal*.

15. This is the term used by L. Westphal and C. Dahlman, "The Acquisition of Technological Mastery in Industry," paper presented to U.S.-China Conference on Alternative Strategies for Economic Development, Wingspread Conference Center, Wisconsin, 1980.

so on. In some cases—for example, agriculture—the imported technology may be unusable without major adaptation; in others it can be used but only very inefficiently. The more sophisticated the technology, the greater the need for technological resources, for effective adaptation and use.

There are large differences in factor productivity between the same industries in different countries.[16] Even with identical technologies, marked differences are apparent.[17] In part, these differences are due to differences in infrastructure, in managerial ability, and in the "quality" of the workforce. But they are also partly due to differences in the extent to which the technology is adapted to the circumstances in which it is used: the greater the differences in conditions between the situation for which the technology is designed and the situation in use, the greater the likely difference in factor productivity. By adaptation to the new conditions, imported technology can be made more efficient. Maxwell's study of the steel firm in Argentina shows how a succession of technological adaptations can raise output and productivity substantially.[18] In countries where

16. As indicated by the evidence collected to test the "Hirschman hypothesis." See S. Teitel, "Productivity, Mechanisation and Skills: A Test of the Hirschman hypothesis for Latin American Industry," *World Development*, 9 (1981).

17. See the comparison of cement production in Indonesia and Florida by L. Doyle, *Inter-Economy Comparisons—A Case Study* (Berkeley: University of California Press, 1965).

18. A similar example of a steel firm in Brazil is contained in C. S. Dahlman and T.V. Fonseca, "From Technological Dependence to Technological Development: The Case of the Usiminos Steel Plant in Brazil," IDP/ECLA/UNDP/IDRC. Regional Program of Studies on Scientific and Technical Development in Latin America, Working Paper No. 21, 1978.

resources are devoted to technological adaptation, rapid increases in productivity occur on the basis of imported technology.

In the first half of the twentieth century both the USSR and Japan industrialized on the basis of adapting imported technology. Granick's study of metal-fabricating in the USSR emphasizes the large role of local adaptation to the different, more labor-abundant conditions of the country. Japanese development in the twentieth century has been characterized by heavy use of Western technology. But Japan has also expended considerable resources on adapting this technology. One-third of R&D has been devoted to adapting foreign technology, and the average expenditure devoted to adapting a unit of technology has been greater than the average expenditure on creating a unit of local technology.[19] As a result of labor-using innovations in the thirties in cotton spinning, the machines were run at rates and speeds substantially above those abroad, where they were initially designed. The spindles were adapted to use the coarser yarn widely prevalent.[20] Labor-using innovations have also been noted in Taiwan in textiles, electronics, and asparagus production.[21] Much of South Korea's industrial success has been attributed to its own technological capacity—to

19. T. Ozawa, *Imitation, Innovation and Trade: A Study of Foreign Licensing Operations in Japan* (Ph.D. dissertation, Columbia University, 1966). See also D. Granick, *Soviet Metal Fabricating and Economic Development* (Madison: University of Wisconsin Press, 1967).

20. G. Ranis, "Industrial Sector Labour Absorption," *Economic Development and Cultural Change*, 21 (1973).

21. Ibid.

adapt and diffuse the imported technology.[22]

The apparent existence of a "technology shelf"—an array of technologies available from the developed countries—might suggest that there is no need to create an indigenous technological capacity. But more in-depth investigation of particular processes all point in the same direction: to choose and use an imported technology efficiently requires considerable local technological capacity. Where this capacity is weak, imported technology tends to have low productivity and there are few improvements over time.[23] Thus a strong case can be made for creating local technological capacity simply to operate the imported technology efficiently.

Whether a capacity to assimilate and adapt imported technology is equivalent to "the generation of technology by LDCs" is in part a matter of definition. From a policy point of view, then, it is not so much a matter of formal R&D as of generating the right type of manpower and the right set of incentives for this type of technological change. However, it should be noted that in some countries—as in Japan—adaptation of foreign technology has been a formal part of government R&D policy, as well as being the product of shop-floor activities.

22. See J. Enos, "The Adoption and Diffusion of Imported Techniques in S. Korea," paper for Science and Technology Workshop, Queen Elizabeth House, Oxford, 1980; L. Westphal, Y. Rhee, and G. Pursell, "Foreign Influences on Korean Industrial Development," *Oxford Bulletin of Economics and Statistics*, 41 (1979).

23. M. Bell, D. Scott-Kemmis, and W. Satyarakwit, "Learning and Technical Change in the Development of Manufacturing Industry: A Case Study of a Permanently Infant Enterprise," mimeo, Science Policy Research Unit, Sussex, 1980.

APPROPRIATE TECHNOLOGY

A technology is designed for use in a specific environment. It therefore tends to have characteristics broadly in line with the circumstances of that environment—in particular with its resource availability, the rest of the technology in use, and its organizational features. In addition the technology is designed to produce a particular product, whose characteristics are acceptable in the market for which it is intended.

1. Factor Use. DC technologies normally involve large amounts of capital per worker, in relation to the savings availability in LDCs. With incomes per head five to ten times those in LDCs, there is a correspondingly greater availability of savings per head for social as well as productive investment. In addition, technologies designed in advanced countries often require levels of skill and education of the workforce, which are widely prevalent in developed countries, or can be readily acquired, but which are in very short supply in poor countries.

2. Scale. The typical scale of a productive unit in developed countries has grown enormously over the course of industrial development. Most modern industries exhibit very marked economies of scale. The minimum efficient scale in many industries in many cases is large in relation to the markets of developed countries. For example, one investigation[24] found that in half of manufacturing industry, the minimum efficient scale exceeded half the total

24. C. Pratten, *Economies of Scale in Manufacturing Industry* (Cambridge: Cambridge University Press, 1971). Of course, much depends on how one defines the "minimum efficient" scale, which would depend in part on country specific factors, such as factor prices and transport costs.

TABLE 1
MANUFACTURING AS A PROPORTION OF THE U.K. MARKET, 1976

COUNTRY	VALUE-ADDED IN MANUFACTURING AS PERCENTAGE OF U.K.
Malawi	0.1
Tunisia	0.7
Kenya	1.0
Ecuador	1.3
Singapore	2.0
Nigeria	3.2
Peru	4.1
Philippines	6.6
India	25.4
Brazil	28.2

SOURCE: *World Development Report 1980*, World Bank. Table 6.

market of the United Kingdom. But the market size of the United Kingdom is very large in comparison with that of most developing counties, as Table 1 shows. If high-scale plant designs are introduced into LDCs, a monopolistic structure of production is often unavoidable, while the few plants that are in operation suffer from excess capacity unless they are able to exploit export markets.[25] In the early stages of production this is often not possible.

3. Infrastructure. Technologies are designed to work efficiently with a particular infrastructure. This includes transport, water, and electricity supplies. It also includes services such as accountancy, legal and government services, insurance, and so on. Without the type of infrastructure for which it is designed, the technology may operate very inefficiently or not at all. Thus having introduced advanced-country technology, LDCs are constrained also to try to reproduce the infrastructure, even though from

25. See M. Merhav, *Technological Dependence, Monopoly and Growth* (New York: Pergamon, 1969), whose analysis of the impact of advanced country technology on LDCs is based on the scale issue.

other points of view this might not be the best use of resources. For example, securing the standards of transport required by the modern technology may mean leaving little to spend on rural transport.

4. Products. Products incorporate a bundle of characteristics[26]; these characteristics are designed for the tastes and income levels of the main consumers for whom they are intended. An important aspect of increasing income is improvements in products to meet given needs: for example, private cars replace public transport. But in achieving these improvements cost increases are normally involved. Many advanced-country products are inappropriate for poor countries, incorporating standards well in excess of those the average household can afford.

There is a strong case for appropriate technology—that is, technology designed for the needs and resources of low-income countries. Such an appropriate technology would be more labor-intensive than advanced-country technology, less

26. See K. Lancaster, "A New Approach to Consumer Theory," *Journal of Political Economy*, LXXIV (1966).

skill-intensive, on a smaller scale, and produce simpler and more appropriate products.[27]

Broadly, one may distinguish two types of appropriate technology: appropriate technology for the modern sector, which consists of the adaptation of modern-sector, advanced-country technologies in more labor-intensive directions; and appropriate technology for the traditional sector, which upgrades and improves traditional technologies. Both have been neglected in terms of information collection and dissemination, research, and development. Promotion of more appropriate technology in both categories may involve improvements in information dissemination, minor or major modification of existing techniques, or the development of entirely new techniques and products. For completeness perhaps a third category should be included: more appropriate systems of public services, including more appropriate products and delivery systems.

There are examples of each type of innovation. In the discussion of technological mastery, we cited examples where foreign technology had been adapted to fit local circumstances—in particular different factor availability and different availability of raw materials. The net result was that the overall capital/labor ratio in textiles in Japan, in the early part of this century, and in Taiwan and South Korea was much lower than the current ratio in advanced countries.[28] In South Korea an old vintage technology—a semiautomatic loom—has been adapted to meet Korean circumstances so that it is more labor-intensive and also lower-cost than the imported—automatic loom—technology.[29] In India, a tractor has been developed with capital costs half those of the imported alternative. An indigenously designed casting process involved investment of Rupees (Rs.) 1000 compared with Rs. 500,000 for imported equipment.

New technologies have also been designed for use in the small-scale and traditional sector, for example, the small-scale open pan sulphitation technique for sugar processing.[30] Considerable research has been devoted to biogas plants for small-scale and economical production of energy. Other examples include solar heating devices and improvements in cooking stoves.[31] While it is not difficult to find a few examples, the scale of activities to develop appropriate technologies is very small compared with activities in advanced countries in developing modern technologies.

In order to promote appropriate technology on any significant scale it is essential to develop efficient alternatives to the sort of technologies designed in advanced countries. Thus a major R&D effort is required, as well as the relevant changes in selection mechanisms, so that appropriate technology is selected. But not all LDC technological activity necessarily produces appropriate technology. For this it

27. See F. Stewart, *Technology and Underdevelopment* (New York: Macmillan, 1977), ch. 4.

28. See Ranis.

29. See Y. Rhee and L. Westpha, "A Micro Econometric Investigation of Choice of Technology," *Journal of Development Economics*, 4 (1977).

30. See M. K. Garg, *Mini Sugar Project Proposal and Feasibility Report* ATDA; Lucknow, 1979.

31. See examples in G. McRobie, *Small is Possible* (Jonathon Cape, 1981); and M. Carr, *Economically Appropriate Technologies for Developing Countries* (I.T. Publications, 1976).

is also necessary to have an incentive system that favors this type of research and a distribution of R&D activity much more concentrated on the small-scale and traditional sector. Moreover, while there has been a fair amount of adaptive research, there has been little in the way of product innovation.[32] This is partly a question of motivation of R&D but also of the nature of markets— which tend to be weighted toward elitist consumption patterns and Western products.

INDEPENDENCE

So long as LDCs lack the capacity to create their own technological innovation, they remain dependent on the advanced countries for technology. This in itself is obviously an aspect of dependence. Moreover, it puts such countries in a weak bargaining position for the acquisition of technology. It also forms an important—perhaps critical—role in perpetuating the more general dependent relationship between North and South.

Critics of the "dependency" view of underdevelopment argue that the concept lacks meaning, or, where it is admitted to have meaning, it lacks operational relevance.[33] All countries with any economic relationships with other countries are in some sense dependent, it is argued. As far as technology is concerned, many developed countries import a very large proportion of their technology, while all developed countries import some of their

technology. Does this make DCs also "dependent"? And if it does, does it matter?

There is obviously some force in these criticisms. Certainly, dependence or indeed technological dependence cannot be precisely defined, nor can the obverse, technological independence, since few would wish to argue that a country is only independent if it is autarchic; hence the dividing line distinguishing dependence and independence is unavoidably arbitrary. Nonetheless, acccepting that the precise dividing line is arbitrary, there is obviously a difference between a country that imports most of its technology and lacks the capacity to create its own, and a country that enjoys a two-way trade in technology and could rely on its own technological resources for a large part, say over half, of its technology. One rather mundane but relevant obvious difference is that a country that imports all its technology will have to pay a heavy cost in foreign exchange.[34]

The United Nations Conference on Trade and Development (UNCTAD) estimates for 1968 were that LDCs were paying 5 percent of non-oil exports earnings for their technology imports, and these were expected to rise by 20 percent per annum.[35] These are the overt costs, but there is also a set of hidden costs. These hidden costs include inflated prices for imports, or overinvoicing, and underinvoicing of exports. These are the direct costs,

32. In South Korea, Westphal et al. found very little indigenous product innovation in comparison with process innovation. See Westphal, Rhee, and Pursell.

33. See S. Lall, "Is 'Dependence' a Useful Concept in Analysing Underdevelopment?", *World Development*, 3 (1975) for an illuminating critical discussion of the concept.

34. But these, although high, may be much lower than the cost of reproducing the technology in the country, as argued earlier.

35. UNCTAD, "Major Issues Arising from the Transfer of Technology to Developing Countries," TD/B/A.C.11/19 Rev.1, 1975.

which have been estimated to be quite high in some industries.[36]

There are also important indirect costs that take the form of restrictions on various forms of activity. These restrictions include tied inputs, restrictions—and even bans—on exports, limitations on competing supplies, and constraints on the dynamic effects of transfer by, for example, limiting local R&D.[37]

The market for technology is an area where bargaining necessarily plays an important role. Once a particular technology has been developed the marginal cost of transferring it to others is low. Therefore with competitive prices, the prices of technology would be very low—often too low to justify its initial development. Consequently all kinds of legal and informal restrictions have been introduced to enable those who develop technology to recoup costs on it by charging more than the competitive price. The upper limit to the price is set by the cost of the purchaser reproducing the technology himself. This upper limit is higher the less the capacity of the purchaser to do so. In the extreme case, lacking technological capacity, the costs of reproducing the technology by the purchaser may be infinitely high, but in this case an upper limit is set by the value of the technology to the recipient. Because of the large gap between the competitive cost and the maximum price a buyer would

be prepared to pay, there is considerable scope for bargaining. For the buyer the bargain is a difficult one because he does not know precisely what he is getting—if he did there would be no need to acquire it. But the capacity to bargain roughly depends on how much the buyer *does* know about the technology, about alternative sources of it, and how easily he could reproduce the technology himself. All these factors depend on the technological accumulation in the country concerned— that is, its technological capacity. Thus the greater the technological dependence, the worse a bargain likely to be struck and the higher the cost to the purchaser.

Technological dependence also supports a generally dependent relationship between countries. With technological dependence often goes (and not just coincidentally) dependence for managers and parts, and foreign ownership or equity participation. In some cases this is because the technology seller insists on these other aspects as part of the bargain. For example, IBM will supply technology only if it also owns the plant. In other cases the countries themselves find the idea of getting the package as a whole attractive, in comparison with the effort required in putting the package together for themselves.

Only countries with tough governments determined to keep their relationships at arms' length— as, for example, historical Japan or the USSR—are able completely to disentangle the technology from the rest of the package and thus permit a certain degree of independence in their relations with the countries who supply their technology. But wherever technological dependence is nearly complete, a country is inevitably dependent on its technology

36. See C. V. Vaitsos, *Intercountry Income Distribution and Transnational Enterprises* (New York: Oxford University Press, 1974), and P. V. Roumelotis and C. P. Golemis, "Transfer Pricing and the Power of Transnational Enterprises in Greece," in *Multinationals and the Market: Transfer Pricing and its Control,* ed. R. Murray (Harvester, 1981).

37. UNCTAD, "Major Issues."

suppliers to such an extent that this can prevent the country taking independent action in other spheres. An example from the past is that of the Suez Canal. So long as the Egyptians believed, as they were frequently told, that they could not possibly run the canal themselves, their whole economic and foreign policy was constrained by that fact. Similarly, the nationalization of copper in Zambia did little for independence so long as technology dependence remained.[38]

CONCLUSION

The main accepted economic theories of resource allocation—the theory of comparative advantage and that of cost-benefit analysis—appear to suggest that LDCs should continue to rely on the advanced countries for their technology. But these arguments do not allow, for three important reasons, for LDCs to create their own technological capacity. First, even to operate an imported technology efficiently requires considerable technological adaptation. Second, technology from advanced countries tends to have increasingly inappropriate characteristics. To create appropriate technology it is necessary to develop technologies in LDCs. Finally, technological dependence involves quite heavy costs, as well as affecting the whole relationship between DCs and LDCs in such a way as to perpetuate dependence. The hewers of wood and the drawers of water—especially if they rely on others to tell them how to hew the wood and draw the water—are unlikely to achieve equality with

their technological masters. The arguments for independent technology generation have been well summarized by Rosenberg:

Of course it is of enormous benefit to them [the LDCs] to be able to import this equipment, even where the equipment is not optimally factor-biased. But if new techniques are regularly transferred from industrial countries, how will the learning process in the design and the production of capital goods take place? Reliance on borrowed technology perpetuates a posture of dependency and passivity. It deprives a country of the development of precisely those skills which are needed if she is to design and construct capital goods that are properly adapted to her own needs. What, then, are the prospects for underdeveloped countries ever becoming efficient producers of capital goods and, in particular, developing a technology with factor-saving biases more appropriate to their own factor endowments? In the past, as I have argued, the appropriate skills were acquired through an intimate association between the user and the producer of capital goods. In the absence of these experiences, what substitute mechanisms or institutions can be established to provide the necessary skills?[39]

The various reasons for LDC capacity to generate technology contain implications for the type of technology to be generated. For technological mastery type of technology, it is essential that technical innovation in LDCs be firmly rooted in the productive activities, not located at a distance in research laboratories. The same is true for the creation of appropriate technology. Technology will be appropriate only if it reflects the environment in which it is to be operated. But to create appropriate technology requires more technological resour-

38. See C. Harvey in *Economic Independence in Africa*, ed. D. Ghai (East African Literature Bureau, 1973).

39. N. Rosenberg, *Perspectives on Technology* (Cambridge: Cambridge University Press, 1976).

ces to be devoted to the small-scale and traditional activities than would be justified by the technological mastery argument alone. The technological independence issue raises difficult questions for the form of technology generation. Does technology which is primarily based on imported technology—as, for example, in Japan—create independence as much as technology which is independently created? Would the latter be as efficient as the former? Any effort to recreate technologies known abroad must inevitably be a waste of resources. There are no simple answers, although it is clear that the independence-type arguments require local technological capacity and an arms'-length relationship with foreign technology suppliers.

Some general conclusions can be drawn for technology generation in LDCs. First, the existence of technically qualified manpower is an essential prerequisite. Second, old style formal R&D in laboratories is not by itself effective; research must have strong links with productive activity. Third, no country can aim for complete technological independence; trade in technology will continue to be an essential aspect of economic development.

ANNALS, *AAPSS*, **458**, November 1981

Technology Transfer to Developing Countries: Implications of International Regulation

By RACHEL McCULLOCH

ABSTRACT: Technology imports are central to the economic performance and development prospects of poor nations. However, while imported technology has helped some nations to achieve rapid industrialization, critics have pointed to a host of actual and potential abuses in the laissez-faire transfer process. As early as the 1960s, developing countries began to adopt national policies regulating technological imports, with particular attention to transactions with Northern multinational corporations. Some of the same nations have led a drive within the United Nations Conference on Trade and Development to establish an international code of conduct governing North-South technology transfer. This article examines motives for and probable consequences of an international code, focusing on the implications of international policies toward technology transfer for the effectiveness of existing national regulation.

Rachel McCulloch earned an undergraduate degree at the University of Pennsylvania and master's and doctoral degrees at the University of Chicago. A former faculty member at the University of Chicago and Harvard University, she is now associate professor of economics at the University of Wisconsin—Madison and a member of the Advisory Council of the Office of Technology Assessment, U.S. Congress. She is the author of many articles and monographs on international economic theory and policy.

ACCESS to advanced technology is central to the economic performance and development prospects of poor nations. Developing nations gain access to advanced technology mainly through technology transfer, that is, by "importing" the fruits of successful foreign research and development efforts. For industrial technology these transfers are primarily commercial transactions. Suppliers, typically Northern-based multinational corporations (MNCs), most often transfer technology as an integral part of a foreign direct investment "package," but also through licensing of foreign production, management and construction contracts, or exports of capital equipment.

North-South technology transfers have usually been initiated by the Northern suppliers, with direct payments and other contractual terms set on an essentially laissez-faire basis. However, as early as the 1960s, a number of developing countries began to adopt national policies regulating technological imports, with particular attention to transactions with Northern multinational corporations. During the 1970s, some of the same nations have led a drive within the United Nations Conference on Trade and Development (UNCTAD) to establish an international code of conduct governing North-South technology transfer. This article examines motives for and probable consequences of an international code, focusing on the implications of international policies toward technology transfer for the effectiveness of existing national regulation.

TECHNOLOGY TRANSFER AND THE NIEO

Demands of the South for a New International Economic Order (NIEO) have deep political roots as well as strong economic motives. Participation in decision making at the global level is demanded not only as a means of safeguarding Southern economic interests, but also as a confirmation by the North of the equal rights of developing nations as members of the international community.

Proposed international regulation of North-South technology transfer reflects Southern dissatisfaction with the status quo and the conviction that current institutional arrangements—or the lack of them—work systematically to the disadvantage of Southern transactors. Indeed, although the issue of access to technology predates the NIEO, there is no other area of North-South economic relations that so clearly illustrates the conditions and relationships central to the South's quest for a new order. Nowhere in North-South relations is international economic inequality more apparent, the South's dependence on the North more clearly underscored, or the perceived gains to the South from altering established patterns of North-South transactions more alluring.

In the area of technology transfer, the political stakes are perhaps higher than for any other issue on the NIEO agenda. This is because the technology issue is so dominated by the role of the United States. Even though U.S. research and development expenditures have been declining as a share of national income while those of most other Organization for Economic Cooperation and Development (OECD) nations have increased, the United States still performs more research and development than all other OECD countries combined. Furthermore, a large share of the industrial research and development performed in other OECD nations is actually carried out by local affiliates of U.S. multinational corporations. And, in addition, the direct investment activities of American MNCs constitute by far the most important channel by which industrial technology created in the North is transferred to developing nations. Thus the technology issue might almost be termed an area of U.S.-South conflict.

NATIONAL REGULATION OF TECHNOLOGY TRANSFER

In the 1950s and 1960s, with large numbers of technologically backward nations gaining independence, the technology issue was viewed principally as a straightforward resource-transfer problem: developing nations wanted more technology at lower cost. Given the supposed public-good character of existing technology and the acknowledged existence of market imperfections that could operate systematically to the disadvantage of developing nations with respect to the terms of trade in technology imports, there was both an efficiency and a distributive rationale for calls to facilitate and subsidize technology transfers to the South.

These calls were met partly by a host of bilateral and multilateral programs, mainly under the control of the North, to increase Southern access to advanced technology. However, MNCs acting from private, profit-oriented motives remained the dominant channel for transfers of industrial technology. Thus while the volume and character of North-South transfers were influenced to a degree by the policies of various multipurpose international agencies—especially the World Bank and UNCTAD—the terms of

most transactions were still determined by prevailing market conditions and the relative economic power of the participants. Yet most developing nations saw little to be gained from regulation of technology imports. The rapid expansion of industrial capacity and national income desired by planners appeared to depend critically on access to Northern capital and advanced technology, and individual nations were seen to have little bargaining power in transactions with giant Northern multinationals. Few countries subjected the terms of technology transfer contracts to close scrutiny.

The uncritical rush to import technology brought mixed results. While a number of countries did achieve rapid industrialization by importing Northern technology, critics were able to identify a host of actual and potential abuses in the laissez-faire transfer process. In addition to excessive direct and indirect costs, major issues raised include the following: contractual restrictions that reduce the value of imported technology and interfere with national sovereignty; complex packaging arrangements that tie importers into perennial dependence on Northern suppliers; importation of technologies that are "inappropriate" to economic and social conditions; centralization of research and development functions in the industrial nations, retarding the development of indigenous technological capacity in the importing countries; and the role of the patent system in preventing technology transfer, thus establishing monopoly markets for Northern exporters.[1]

1. See, for example, G.K. Helleiner, "International Technology Issues: Southern Needs and Northern Responses," in eds. Jairam Ramesh and Charles Weiss, Jr.,

These new perceptions encouraged a gradual shift on the part of Southern nations from a passive to an active role in the technology transfer process.[2] Individually—for example, Mexico, Brazil, and India—and in groups—the Andean Pact nations—some of the more advanced among the developing nations began to regulate technology imports. National regulation typically entailed screening of proposed contractual arrangements by a government agency, with an eye toward eliminating excessive direct payments and restrictive terms limiting the options of the acquiring firm in areas such as modification and adaptation of the technology, sources of inputs, marketing of outputs, and use of trademarks.[3]

Evaluation of the results of national regulation inevitably entails a high degree of subjectivity. It is impossible to measure in a meaningful way either the amount of technology transferred or its price. Furthermore, there are several important reasons, apart from national regulation, to expect changes over time in the terms of technology trade. For any specific technology, the number of potential suppliers increases over time as the technology becomes standardized and diffusion and imitation occur. Also, the national "absorptive capacity" that determines the true resource cost of implanting a new

technology improves as a country develops. Both factors should result in more favorable terms for technology importers. On the other hand, the progress of development within a country also means that technological needs shift toward more advanced and more expensive technologies available only from a small number of suppliers. Nevertheless, proponents seem convinced that terms have been improved as a result of negotiations between filtering agencies and technology-exporting firms and that many objectionable restrictive practices previously standard have been eliminated through national surveillance.[4] While in some cases the initial regulations were later deemed excessively stringent and were modified accordingly, most countries that have followed the course of national regulation appear well satisfied with the results.

PROPOSED INTERNATIONAL REGULATION

Most of the nations that have adopted national regulation of technology transfers are now pressing for international regulation along similar lines. The proposed UNCTAD code would separate technology transfers from other trade and investment transactions,

Mobilizing Technology for World Development (New York: Praeger, 1979), pp. 85-97.

2. This trend was reinforced during the 1970s by a dramatically altered appraisal of the potential influence and power of Southern nations in the world economy.

3. For a summary of the terms and effects of national regulation, see Debra Lynn Miller, *Political Struggles for Control of the International Transfer of Technology* (Ph.D. thesis, Department of Government, Harvard University, Sept. 1979), pp. 170-93.

4. Evidence in support of gains from national regulation is cited in Frances Stewart, "International Technology Transfer: Issues and Policy Options," World Bank Staff Working Paper No. 344 (Washington, DC: The World Bank, July 1979), pp. 58-61. However, other investigators have reached opposite conclusions—sometimes on the basis of the same statistics. See, for example, John E.S. Parker, "Pharmaceuticals and Third World Concerns: The Lall Report and the Otago Study," in *The International Supply of Medicines*, ed. Robert B. Helms (Washington, DC: American Enterprise Institute, 1980), pp. 135-46.

emphasizing the Southern view that access to technology is an issue of utmost priority for poor nations. In place of the current laissez-faire system, the South would substitute one based on the principle central to the NIEO program: preferential treatment of Southern nations to promote rapid growth and ensure equity in economic relations with the North.

The UNCTAD international code of conduct on the transfer of technology now under consideration embodies four major objectives of its proponents.[5] Most important, the code would establish the right of technology-importing nations to regulate technology transfer through national law. The code would also curtail the use of restrictive business practices, encourage the sharing of technological information, and promote the development of indigenous technological capabilities of Southern nations.

This program to restructure world markets for technology to meet better the needs of Southern nations rests on two broad underlying principles common to the entire NIEO drive. These principles, here termed "enlightened dependency" and "collective self-reliance," are to some extent complementary in that progress toward the goals of one may be conducive to or even necessary for attainment of the other. However, their immediate implications for action are in conflict, especially with regard to ordering of immediate and longer-term benefits from regulatory action. This same

5. Each of these objectives represents the substance of at least one chapter of the draft code. See United Nations Conference on an International Code of Conduct on the Transfer of Technology, *Draft International Code of Conduct on the Transfer of Technology as of 6 May 1980* (New York: United Nations, 1980), pp. 3-26.

conflict is inherent in the specific proposals outlined in the UNCTAD draft code.

The principle of enlightened dependency emphasizes restructuring of relations with the North, and especially with Northern multinational firms, as a means of enhancing Southern gains. Improved information, international oversight of contractual terms, and coordinated action by Southern nations are measures to improve the bargaining power of technology importers and thus achieve more favorable overall terms.

Collective self-reliance, a term that occurs frequently in NIEO documents, does not have a single clear interpretation. In the context of technology transfer, it refers to proposals for sharing of Southern knowledge resources, the substitution of intra-South transactions for North-South transfers, and the development of indigenous research capacity, especially that aimed at the "appropriate" products and processes not readily available from the Northern technological establishment.

The first principle appears to accept the inevitability of continued dependence on the North as the primary source of technological advance, but seeks to improve the South's capacity to derive the greatest possible benefits from technology transfer transactions. The second principle places more stress on the immediate reduction of dependent relations with the North. A key difference is the implied time horizon. Gains in the short run are almost certain to be greater to the extent that the principle of enlightened dependency is stressed. A collective self-reliance emphasis represents orientation toward a longer-run perspective. Like the older infant-industry arguments once championed by many of the same nations, collective self-reliance stresses vaguely specified dynamic processes to be set in motion by restriction of technology imports from Northern suppliers. Even the most enthusiastic supporters of this principle do not propose to cut the South off entirely from Northern technological advances; nevertheless, high short-run sacrifices are likely to be entailed, so that the appeal of this regime rests on the assumption that eventual gains will be large enough and will come soon enough to outweigh these costs.

At least initially, policies to achieve enlightened dependency would also lead to some reduction in the North-South flow of technology. With terms more favorable to the South, some previously acceptable arrangements would no longer be sufficiently attractive to Northern suppliers. In the longer run, successful development could have more complicated consequences for the degree of self-sufficiency. The technological needs of the poorest nations can be met mainly by highly standardized technologies available from many competing sources, including some in the South. But more advanced developing nations are likely to import newer technologies that are obtainable only from one or a few Northern sources and at higher cost.

If national regulation has indeed been a success, why have the very nations that already benefit from their own national policies also become the most vocal protagonists of international action along similar lines? Success in establishing an international code governing technology transfer would reinforce the leadership roles within UNCTAD of those nations—mainly Latin American—now imposing national regulation in this area. To the extent that the standards set in the international code are patterned after existing national regulation, the effect would be to legitimate national policies—possibly reducing the likelihood of retaliation by technology-exporting countries. Furthermore, international regulation that imposes uniformly stringent national regulation on all importing nations would imply an improved market position for importers as a group, with resulting collective gains in the form of more favorable terms.

The sections that follow provide a closer look at the economic characteristics of North-South technology transfer and the likely consequences of regulation along the lines of the proposed UNCTAD code.

WHAT DOES TECHNOLOGY TRANSFER TRANSFER?

While technological imports are a fact of modern economic life for all nations, including the most advanced, the precise meaning of technology transfer depends critically on the stage of development of the importing nation. Much of the debate on the issue of North-South technology transfer centers on the notion that knowledge is a "public good"—that is, that once in existence, it can be extended to additional users at negligible social cost. The word "technology" suggests blueprints or formulas that, once obtained, guarantee more or better outputs from given productive inputs. This public good image is accurate to a degree for intra-North technology trade, especially in some advanced technology. However, the technology gap that separates North from South is overwhelmingly a difference in available "know-how"—that is, practical and conceptual skills embodied in the national labor force of a society. For Southern technology imports, it is often appropriate to substitute the term "know-how" for technology. In technology transfer to the South, transformation of inputs, especially labor, is a vital part of the process.

Of the body of economically useful knowledge that supports modern industrial production, only a small fraction is protected by patents or even trade secrets. Most is in the public domain. Transfer of know-how technology is largely a question of training and teaching, sometimes accomplished through formal educational programs and international exchanges, but usually through informal learning-by-doing and on-the-job training. This point is critical to the debate over North-South technology transfer because while knowledge in the sense of blueprints or formulas is indeed essentially a public good and can be made available to additional users at a marginal social cost near zero, the same is not true for training and teaching, activities that absorb valuable resources. Where a high component of know-how is essential to consummating the transfer, the social cost of extending modern technology is likely to be significant.

That training and skill acquisition are at the heart of the technology transfer issue is corroborated in a number of ways. First, the key role of labor-force education levels has been widely acknowledged in the literature of economic development; the link of education to diffusion rates for new technology within a given country has been established empirically. Second, both the rhetoric and the policy focus of national regulation have been to increase the extent to which nationals are trained to participate in every stage of local production processes employing imported technologies. While this requirement is usually embedded in calls for "increased national sovereignty" and "breaking of neocolonial ties," the central issue at the practical level is how many nationals will be trained to do what. Third, the preferred Southern term for the "brain drain"—emigration of skilled workers and professionals—is "reverse transfer of technology." This is apt nomenclature; access to modern technology benefits a developing nation only to the extent that some *person* has the knowledge to implement it. A fine library of engineering texts cannot substitute for one moderately skilled engineer.

Finally, it is suggestive that the rate of expropriations of foreign subsidiaries is lower for manufacturing than for extractive industries and, within manufacturing, falls with the level of technological sophistication.[6] When the skills of managers—often foreign nationals—are the essential ingredient of an enterprise, nationalization can kill the goose that lays the golden eggs. This relationship underlies in part both the reluctance of some multinational firms to train nationals in all aspects of local operations and the trend away from foreign direct investment as a technology transfer vehicle toward management contracts that implicitly or explicitly recognize the limited time horizon until an adequate national skill base for independent national operations has been established.

6. Stephen Kobrin, "Foreign Enterprise and Forced Divestment in LDCs," *International Organization*, 34(1):65-88 (Winter 1980).

Recent national regulation and proposed international regulation aimed at reducing contractual restrictions on the application, extension, and adaptation of imported technology can be seen as ensuring legal rights to control over technology; skill acquisition remains a necessary condition for effective exercise of these rights in the national interest. Thus while skill acquisition is undoubtedly central to North-South technology transfer, legal factors can still play a key role in determining national benefits from the transfers.

THE PATENT SYSTEM AND SOUTHERN NEEDS

The production of new knowledge, as with other types of public goods, presents a conflict between static and dynamic efficiency in a market economy. For knowledge already in existence, static efficiency is best served by making it freely available to all potential users. This is roughly the system used in pure science, where tax revenues of the major industrial economies supply the lion's share of resources for innovation. For commercially useful knowledge, the patent system represents capitalist society's compromise between static and dynamic efficiency. By granting innovators a temporary monopoly over the fruits of their efforts, market economies provide a private incentive to future innovative activity and, more significantly, to prompt disclosure of useful findings.

The role of patents in protecting innovators' priority has been steadily declining. Legal costs of obtaining patents and especially of defending patent rights are high. Furthermore, the degree of disclosure entailed in obtaining a patent can facilitate imitation by commercial rivals. For these reasons industrial innovators have resorted increasingly to "trade secrets" and intentionally heightened complexity to improve the "appropriability" of new products and processes.[7]

The role of patents in international technology transfer has been declining steadily for decades; it is minor in the case of North-South technology transfers because the bulk of relevant technology is in the public domain. Thus the international agreement governing the transfer of patented technology, known as the Paris Convention, is of small practical significance to the South. It is, however, of considerable symbolic importance; some regard it as clear evidence that current international arrangements favor the interests of the rich technology-exporting nations.[8] The most crucial aspect of the Paris Convention for Southern nations is that member states agree to provide "national"—that is, nondiscriminatory—treatment of foreign patentees under their own patent laws. Thus special treatment for poor nations and their innovative sectors is precluded.

Developing nations argue that the Paris Convention is aimed primarily at protecting the innovator and that poor nations are not well served by adherence to its Northern-inspired provisions. Developing nations are overwhelmingly users of knowledge created by others, so that

7. Stephen P. Magee, "Information and Multinational Corporations: An Appropriability Theory of Direct Foreign Investment," in *The New International Economic Order: The North-South Debate*, Jagdish N. Bhagwati, ed. (Cambridge: MIT Press, 1977), pp. 317-40.

8. Miller, pp. 81-115.

protection of mainly Northern innovators is a minor motive for them. Furthermore, Northern innovations are often deemed "inappropriate" to the needs and conditions of Southern nations. In any case, the collective contribution of Southern royalties to the total market reward for innovation is small and thus has little influence on the overall profitability of past research and development efforts or expected profits from future innovation. For these reasons, some critics of the current patent regime argue that Southern interests would be better served by a legal framework offering less patent protection for innovations of Northern origin.[9]

The case for reduced patent protection of technology from Northern sources overlooks several important considerations. First, sales to developing nations of patent-protected products are large and growing. A Southern legal framework offering little protection for innovations originating in the North would reduce royalty payments to affected Northern exporters and thus could have a much larger effect on the total profitability of innovation than data on South-North royalty payments alone would suggest.

Second, Northern technology is largely the fruit of the research and development efforts of Northern multinational corporations, planned and carried in reference to the markets and production conditions of the North. Thus, almost by definition, it is likely to be in some degree "inappropriate" to the South. However, to the extent that past research and development represents sunk

costs and that existing technology can be applied in the South at a relatively low additional cost—not always a sensible assumption, for reasons discussed in the previous section—there is a tradeoff between further research and development costs to tailor products or processes to Southern users and higher social and economic costs of using available but inappropriate technology without modification. In at least some cases "inappropriate" technology may be the more appropriate social and economic choice.[10]

Furthermore, while little current Northern research and development is aimed specifically at products and processes tailored to Southern markets and production conditions, growth of Southern per capita incomes and ongoing industrial transformation is likely to make innovation of this kind more profitable in the future. But only if innovators have assurance that they can capture some of these potential profits will Northern research and development oriented toward Southern needs expand.[11]

Still more important than these considerations is that the model of knowledge as a public good with a transfer cost near zero is largely irrelevant for North-South technology transfer. For these transactions the major costs of making the transfer lie in creating required local

9. For example, see Constantine V. Vaitsos, "The Revision of the International Patent System: Legal Considerations for a Third World Position," *World Development*, 4(2):85-102 (Feb. 1976).

10. National policies in developing nations sometimes actually promote the importation of inappropriate technology. Capital subsidies, restrictions on imports of used machinery, and insistence on the latest (and usually most capital-intensive) technologies all contribute to technology choices inappropriate to local production conditions. High tariffs or restrictive quotas intended to discourage imports of luxury consumer goods create profit incentives for their local production.

11. Parker, pp. 139-40.

know-how and infrastructure. The ineffectiveness of compulsory licensing as a remedy for nonworking of Southern patents suggests that even making patented technology freely available to Southern users would have little immediate impact on its local application by Southern firms.[12]

MARKET IMPERFECTIONS AND THE CASE FOR INTERNATIONAL REGULATION

Many of the more sophisticated arguments in favor of NIEO proposals center on "market imperfections" affecting North-South transactions.[13] In the presence of these market imperfections, standard neoclassical analyses favoring nonintervention on economic efficiency grounds are invalid. NIEO advocates argue also that the nature of these market imperfections not only leads to a smaller global "pie"— that is, departures from allocative efficiency—but also systematically produces a skimpier Southern slice. In the case of technology transfer, three interrelated market imperfections are identified: incomplete information for buyers, lack of competition among sellers, and restrictive business practices routinely included in contractual arrangements.

Lack of adequate information is the most important of the market imperfections and contributes to the other two. A heterogeneous and constantly changing technology base makes information costly to obtain;

12. Miller, pp. 96-97.
13. See, for example, Stewart p. 101, and G.K. Helleiner, "World Market Imperfections and the Developing Countries," in *Policy Alternatives for a New International Order*, William R. Cline, ed. (New York: Praeger, 1979), pp. 357-89.

for most developing nations, the technical expertise necessary to acquire and evaluate information from alternative suppliers is absent. Scarcity of information and the skills required to assess its relevance lead decision makers to choices that are not always in the best interests of technology-importing nations. A key role in these choices is frequently played by Northern experts supplied by international organizations such as the World Bank; the value judgments, often implicit, of these experts are likely to differ from those of national policymakers.

The second market imperfection argument for national regulation is built on the assumption that the number of potential suppliers is quite small, sometimes just one. Thus there is no assurance to the buyer that competition among sellers will afford protection from excessive costs or disadvantageous contractual terms. The small perceived number of suppliers is partly a reflection of information scarcity. Buyers are often unaware of alternative sources of similar technology, while most potential suppliers are likely to have familiarity with market and production conditions for only a handful of potential importers of their technology.

Even where monopoly rights to a product or process are protected by patents, similar technology is usually available elsewhere. For most of the technology required by developing nations, the closing of the technology gap between the United States and other industrial nations has greatly increased the number of competing suppliers. Only in the case of the newest advances is supply likely to be monopolized by a single firm. Still, the problem of information may imply that competition among suppliers will be slight, so that buyers cannot rely on market forces in obtaining the best possible terms.

A third argument is that absence of competition among sellers and lack of adequate information on the part of buyers allows suppliers to impose a wide range of restrictive business practices on technology users as part of standard contractual arrangements. The restrictive business practices—as many as 40 have been identified in some UNCTAD documents[14]—prevent Southern nations from extracting the maximum possible value from imported technology. Examples include tie-in arrangements for procurement of inputs, restrictions on export sales, and assignment of rights to modifications or improvements. Some of the practices would, if employed in domestic contracts, be clear violations of the antitrust laws of the United States and other advanced nations. In many instances, however, and especially where the technology importer is a subsidiary of a multinational corporation, the restrictive practices may conflict with national policy objectives in the importing nation but do not constitute violations of Northern antitrust principles—in particular, these do not require parts of the same corporation to compete actively with one another. In these cases, there is no necessary conflict between the so-called restrictive practices and worldwide allocative efficiency.

Where market imperfections are present, a case for international regulation can be made on grounds of improving global allocative effi-

ciency.[15] Significantly, potential gains from action to correct the market imperfections identified here are likely to accrue largely to those nations with modest internal resources—those without the indigenous skills required to collect and evaluate relevant information, countries that are largely passive in technology transfer transactions.

Of the imperfections affecting technology transfer, lack of adequate information is central. Unfortunately, improved information in a usable form does not appear to be a likely outcome of new international action. The UN system already produces an overwhelming volume of information and documentation relating to the transfer of technology, yet this is apparently inadequate to meet the actual needs of national decision makers.[16] Because the state of technology changes so rapidly, collection and updating of information on what is available and from which sources must be an ongoing and very costly procedure. Probably more important, what is "available" to one technology-importing country is not necessarily available to another. Differences in market conditions and economic infrastructure mean that technology deals are best thought of as made-to-order, rather than selected from a UN-produced catalog.

On the other hand, sharing of information about other nations' contractual arrangements in the recent past could prove helpful by

14. Werner J. Feld, *Multinational Corporations and U.N. Politics: The Quest for Codes of Conduct* (New York: Pergamon Press, 1980), p. 83.

15. This does not, however, imply that regulation will necessarily produce an improvement in allocative efficiency, as some writers appear to assume.

16. One of the few concrete results of the 1979 United Nations Conference on Science and Technology was agreement to set up a global information system for use by technology-importing nations.

identifying the range of terms technology suppliers may be willing to accept. However, this will be useful only for countries with adequate access to basic legal, accounting, and engineering skills. Without these resources at the national level, additional information can have little impact on the quality of decision making.

INTERNATIONAL REGULATION AND MONOPSONY POWER

Although correction of market imperfections is a staple of NIEO rhetoric, the most important economic motive for international regulation of North-South technology transfer lies in a different direction. When developing nations are able to compete with one another through incentives offered to MNCs siting new operations, technology-supplying firms gain at the expense of the importing nations. By acting in concert, Southern nations can lessen such competition among themselves and thus raise their own collective gains—that is, improve the Southern terms of technology trade via a kind of "optimum tariff" strategy. International regulation that produces uniformly stringent national regulation implies an improved bargaining position for the buyers through the exercise of monopsony power.[17]

17. Because lessened competition among buyers would, in itself, constitute a move away from conditions of "perfect competition," it is possible that such collective action on the part of the South could reduce worldwide allocative efficiency, as some Northern critics suggest. However, in the presence of many other kinds of market imperfections, the principle of second best must be invoked; one could reasonably expect an improvement in global efficiency along with increased Southern benefits on second-best grounds.

As in the case of cartel arrangements, this kind of regulatory coordination would benefit some importing nations more than others and could also raise some problems of enforcing the desired uniformity. To the extent that all importing countries must agree to similar terms, it is those countries with the greatest absorptive capacity—the most attractive investment environments and hence usually the nations that are already more industrialized or those with valuable natural resources—that will attract the lion's share of new technology deals. For countries already regulating investment and technology imports, an international code that promotes restrictions elsewhere will be beneficial. However, those benefits will come at the cost of countries that might otherwise receive investments deflected by the stringent regulations of the first group. Thus the less competitive countries will gain only if improved terms on remaining investments can offset the costs from losing others.[18]

Further gains to the more technologically advanced Southern nations would accrue to the extent that the new regime actually promoted collective self-reliance. Such an emphasis would benefit those Southern nations already best able to claim indigenous technological capacity and to accommodate increased demands for services that substitute for Northern technology exports.

A LOOK AHEAD

Work on the proposed UNCTAD code of conduct on technology

18. The argument is roughly the same as the one suggesting that minorities and unskilled workers as a group may be harmed by a minimum wage.

transfer has been in progress since 1975, but some crucial issues remain outstanding. Of these, the most important is the legal character of the code. From the start, the developing nations have maintained "that an internationally legally binding instrument is the only form capable of effectively regulating the transfer of technology," while the advanced industrial nations have argued that the code should consist merely of "guidelines which are voluntary and legally nonbinding."[19] After fruitless attempts to find a satisfactory solution to the impasse, the negotiators agreed that the code would be adopted initially as an advisory document, but to be reevaluated after four years "with a view to bringing about its universal applicability as a legally binding instrument."[20] The developing nations have emphasized that adoption of the code as a non-legally-binding instrument should not be taken as a final decision on the legal nature of the code.

This article suggests that the proposed international code of conduct on the transfer of technology would benefit primarily the countries most active in promoting its adoption. For countries that have already implemented national regulation, an international code would legiti-

mate their own restrictions and reinforce their market position by encouraging other nations to adopt similar regulation. The gains would accrue mainly to the more advanced among the developing nations and could come, at least in part, at the expense of countries less far along in the development process. These consequences depend on the influence of the code on the policies of individual developing nations and are not likely to be affected by the legal nature of the code.

However, the code is also intended to curb certain restrictive practices of the Northern multinational firms that are the major suppliers of industrial technology to developing nations. Here the legal nature of the code could have an important effect on its influence, if any, on actual transactions. As long as the code is merely advisory, its relevance to the activities of private, profit-seeking corporations is likely to be minimal.

A third aspect of the code is an effort to correct existing market imperfections, mainly reflecting problems of inadequate information. The legal nature of the code could influence the level of resources individual advanced nations—and especially the United States—feel obligated to commit to information-sharing activities. However, as already noted, for the poorest nations the problem of evaluating information is at least as great as that of obtaining it. The likely gains to these nations are therefore minor regardless of the code's legal status.

19. United Nations Conference on an International Code of Conduct on the Transfer of Technology, *The Draft International Code of Conduct on the Transfer of Technology: Major Issues Outstanding* (Geneva: UNCTAD, 1980), p. 19.

20. Ibid., p. 20.

The Appropriability Theory of the Multinational Corporation

By STEPHEN P. MAGEE

ABSTRACT: The appropriability theory of the multinational corporation emphasizes the conflict between innovators and emulators of new technologies. Appropriability is "high," and innovators can protect their profits more easily for sophisticated technologies and on breakthroughs that can be transmitted worldwide through the innovator's own subsidiaries. Conversely, appropriability is "low," and multinationals find it less profitable to create simple technologies and ideas that require market transfer. This theory explains the limited role multinationals have played in the development of simple products and simple production technologies, both of which are important to the developing countries. The appropriability theory also predicts that products in Vernon's product cycle will move to stage II when developed countries start successful emulation of the product and to stage III when developing countries start successful emulation. The profit-maximizing price strategy an innovating multinational should follow is to sell new products at below the monopoly price and slowly cut the price of the product as appropriability mechanisms erode. In the long run, the multinational will be forced to sell at the perfectly competitive price. If the multinational has no long-run profit advantage over other producers, its long-run market shares should approach zero as the perfectly competitive price is approached.

Stephen P. Magee is the Margaret and Eugene McDermott Professor and chairman of the Department of Finance at the University of Texas, Austin. He received his Ph.D. in economics from MIT in 1969 and taught at the University of California, Berkeley, and at the University of Chicago before going to the University of Texas in 1976. He has served as an economist for the White House, the Council of Economic Advisers, and the Brookings Institution. He has also served as associate editor of the Journal of International Economics *and the* Review of Economics and Statistics.

NOTE: The author is indebted to the National Science Foundation, the University of Texas, and the University of Reading for research support.

I T is said that Thomas Edison spent more on legal fees to protect his light bulb than he received in fees and royalties from that invention. A similar problem faces innovating multinational corporations. The process of protecting the returns from innovations has been explored in an appropriability theory of multinational corporation (MNC) behavior.[1]

The appropriability theory suggests that the most important consideration facing innovating multinationals is the possible loss of the technology to rivals and copiers. New ideas are public goods, which means that anyone who can figure out how to use them may do so without reducing the use by others. But unauthorized use of new ideas certainly reduces the profitability for innovators. The more difficult it is to protect the profitability of an innovation, the greater the appropriability problem. When applied to the multinational corporation, the appropriability theory suggests that it is more efficient to transfer high technology worldwide inside firms than through the market because

there is less likelihood of it being copied and stolen by outsiders if it is under the control of a single firm. An innovating firm will invest resources to keep others from copying and stealing the idea. The appropriability theory suggests that mechanisms evolve to prevent the loss of high technology and that these form a central theme that can explain much multinational corporation behavior.

It will explain, for example, why MNCs create very sophisticated technologies rather than simple ones. Sophisticated ideas are hard to copy, while simple ideas are easy to copy. Therefore the appropriability problem is particularly severe for simple ideas. One implication of this hypothesis is that multinational corporations cannot be counted on to create the types of technology that are most useful for the developing countries. The developing countries need two types of technologies: simple production processes and simple products. However, the ability of private firms to capture the returns on these types of ideas is difficult.

The appropriability model generates results that both contrast with and help explain the Vernon product cycle.[2] In contrast, the appropriability model suggests that industry age—and not Vernon's product age—is the key to understanding international technology flows. This idea has been explored in the industry technology cycle.[3] Appropriability considerations complement Vernon's theory in explaining

1. The appropriability of an innovation is higher the larger the innovator's profits relative to the value to society of the innovation. The appropriability theory reviewed in this article is a summary and extension of the following three publications: Stephen P. Magee, "An Appropriability Theory of Direct Foreign Investment," in *The New International Economic Order: The North-South Debate*, ed. J. Bhagwati (Cambridge: MIT Press, 1977), pp. 317-40; "Multinational Corporations, the Industry Technology Cycle and Development," *Journal of World Trade Law*, 11:297-321 (July 1977); and "Application of the Dynamic Limit Pricing Model to the Price of Technology and International Technology Transfer," in *Optimal Policies, Control Theory and Technology Exports*, eds. Karl Brunner and Allan Meltzer (Amsterdam: North-Holland, 1977), pp. 203-224.

2. Raymond Vernon, "International Investment and International Trade in the Product Cycle," *Quarterly Journal of Economics*, 80:190-207 (May 1966).
3. See Magee, "Multinational Corporations."

the length of each stage of his cycle. So long as the innovating firms in an industry maintain their technological lead over emulating firms, the industry will remain young and produce new products. When appropriability mechanisms break down (for example, when industry structure becomes less concentrated), emulators in the United States and abroad reduce the profitability of innovations so that the industry's product line shifts to older, more standardized products.

What price strategy will an innovating multinational pursue in the face of eroding appropriability?[4] Even if the multinational is the only producer early in a product's life, it should sell below the monopoly price—since the monopoly price encourages emulator production, which reduces the present value of future profits by more than is recouped in today's profits—and slowly cut the price as appropriability erodes until it hits the perfectly competitive price. The multinational's product market share hits zero just as the product becomes completely standardized. If the market share is still positive with standardization, the multinational has pursued too low a price strategy— that is, discouraged emulators more than was in its own profit interests.

This article reviews in some detail the three studies summarized. The first major section summarizes the appropriability theory and the industry technology cycle, while the second major section explores the profit-maximizing price strategy that multinationals will follow and how they will behave

in the face of certain widely discussed (North-South type) policy changes.

ROLE OF APPROPRIABILITY IN THE CREATION AND DIFFUSION OF TECHNOLOGY

Consider the creation of technology by private firms. Technology is a durable good in that present resources must be devoted to its creation and its existence results in a stream of future benefits. Technology is also a public good in that once it is created, its use by second parties does not preclude its continued use by the party who discovers it. However, use by second parties does reduce the private return on information created by the first party. This last feature has been labeled the "appropriability problem."[5]

The appropriability theory of multinational corporate behavior is a natural outgrowth of the industrial organization approach to international direct investment developed by Hymer, Vernon, and Caves, as well as the views of Arrow, Demsetz, and Johnson on the creation and appropriability of the returns from private market investments in information.[6] The appro-

4. See Magee, "Application of the Dynamic Limit Pricing Model."

5. See Kenneth Arrow, "Economic Welfare and the Allocation of Resources for Invention," in *The Rate and Direction of Inventive Activity: Economic and Social Factors*—a report of the National Bureau of Economic Research (Princeton, NJ: Princeton University Press, 1962), pp. 353-58.

6. Stephen H. Hymer, *The International Operation of National Firms: A Study of Direct Foreign Investment* (Cambridge: MIT Press, 1976). See also Vernon, "International Investment and International Trade in the Produce Cycle"; Richard Caves, "International Corporations: The Industrial Economics of Foreign Investment," *Economica* 38:1-27 (Feb. 1971); Arrow, "Economic Wel-

priability theory suggests that MNCs are specialists in the production of technology that is less efficient to transmit through markets than within firms; that MNCs produce sophisticated technologies because private returns are higher for these technologies than for simple ones; that the large proportion of skilled labor employed by the multinationals is an outgrowth of the skilled-labor-intensity of the production process for both the creation and the appropriability of the returns from technology; and that the relative abundance of skilled labor in the developed countries dictates that they have a comparative advantage in creating, exporting, and capturing private returns on new technologies.

The theory also suggests that the structure of industry and the creation of technology are jointly determined variables. The presence of a monopoly or oligopoly, *ceteris paribus*, encourages R&D and other investments in innovation because appropriability costs are lower for these industry structures. In turn, a major innovation encourages an increase in optimum firm size, so that industry structure becomes more concentrated. Thus, there is two-way causation between new technology and industry structure.

In the process of creating and carrying each product through its life cycle, multinationals must generate four distinct types of information—that is, new technology: product

creation, product development, creation of the production processes, and creation of the markets. These four types of technology investments are described in the following paragraphs. Then the causes and consequences of appropriability are described. The welfare question of whether private markets or government agencies should produce new technology is examined briefly next, followed by a description of an industry technology cycle, paralleling Vernon's product cycle, which emerges from the appropriability theory.

*Four types of technology
   created by multinationals*

Each product goes through a life cycle with a lot of new information created in the early stages and less created as the product matures. Investments in information must be made for the discovery of new products, for their development, for the creation of their production functions, and for the creation of their markets.

First, investments are required to discover new products. While an increasing proportion of total R&D is done within large corporate organizations, many new ideas are still developed by small independent inventors. It is my impression that invention is not the focus of MNC activity: their R&D efforts are focused on innovation, which encompasses the next three types of information.

The second activity requiring large expenditures on scientists and engineers is in product development—that is, applied research, product specification, and prototypes. Mansfield finds that product development frequently requires five to

---

fare and the Allocation of Resources for Invention"; Harold Demsetz, "Information and Efficiency: Another Viewpoint," *Journal of Law and Economics*, 12:1-22 (Apr. 1969); and Harry G. Johnson, "Multinational Corporations and International Oligopoly: The Non-American Challenge," in *The International Corporation*, ed. C. P. Kindleberger (Cambridge: MIT Press, 1970), pp. 35-56.

ten years for major products.[7] These undertakings are the activities engaged in by multinational corporations at the beginning of each product's life cycle. MNCs develop a comparative advantage in moving products through Vernon's product cycle. MNCs are large because it is more efficient to transfer information on development from product to product within the firm rather than through the market. This explains the tendency of multinational corporations to carry more product lines than national firms.[8] Information on avoiding mistakes is usually more costly to transmit through the market than intrafirm.

The third piece of information required in the product cycle is the creation of the production processes. Economists have traditionally assumed the existence of "production functions"—provided by engineers. However, there is a growing awareness that the creation of the production processes is determined like other processes: by the supply and demand for production technology. Also, factor price structures differ between the developed countries (DCs) and the less-developed countries (LDCs).

Production occurs in LDCs so late in the product life cycle that discounting gives the importance of cheap unskilled LDC labor a small weight to the multinational. Another point is that industry structures become more competitive through the cycle as patents lapse, so that the production process becomes "frozen" or standardized at a more capital-intensive level than may be desirable for LDC production. The reason for the "freeze" is that increased industry competition erodes the private market appropriability of the private returns from developing unskilled-labor-intensive production techniques more fully. Still another reason for failure of MNCs to shift from capital-intensive to unskilled-labor-intensive production in the LDCs is quality control. For example, one firm that shifted from mechanized to hand-labor food canning in an LDC quickly regretted the decision after numerous cases of food poisoning developed.

Fourth, investments in information must be undertaken to create product markets. One interpretation of the multinational corporation is that they act like large retail stores in selling new technologies and new information.[9] Let us develop a framework for this theme. Information is closer to an "experience" good rather than a "search" good.[10] This distinction in the advertising literature explains why large retail stores have become important in prescreening for consumers. For example, if a person wishes to buy a very high-quality consumer product, he is more likely to go to Saks Fifth Avenue than to a bargain-basement discount store. "Experience goods" are those whose value to the purchaser cannot be established upon visual inspection and for whom the brand name of the good or the name of the retailer is an important signaling device. Search goods are those whose physical attributes can be examined and successfully

7. Edwin Mansfield, "Technology and Technological Change," in *Economic Analysis and the Multinational Enterprise*, ed. J. H. Dunning (New York: Praeger, 1974), pp. 147-83.

8. Raymond Vernon, *Sovereignty at Bay* (New York: Basic Books, 1971).

9. Lester Telser, "Comment," *Journal of Law and Economics*, 19:337-40 (1976).

10. Phillip Nelson, "Information and Consumer Behavior," *Journal of Political Economy*, 78:311-29 (Mar./Apr. 1970).

compared with the claims of advertisers before purchase and for whom brand names are less important.

Causes and consequences of appropriability

Finally, there is an important fifth piece of information every multinational firm must have before it embarks on developing a particular product. This is appropriability itself.

There is considerable variation across products and processes in the extent to which a private firm can appropriate the returns from an investment in new information. For complicated ideas and technologies, it is relatively difficult for interlopers to steal the idea. For simple ideas, there is a larger sample of potential entrants who can steal the idea and reduce the returns on investments by an innovator. This is one reason for the lack of R&D in highly competitive industries. Loss of appropriability through time is analogous to depreciation: complicated ideas have slow depreciation rates, while simple ideas have high ones. Differential private appropriability leads to social underinvestments by private firms in simple and unskilled-labor technologies.

Coase points out that externalities can be efficiently handled by private parties—rather than the government—if the legal system clearly establishes property rights.[11] One implication of his article is that governmental intervention is not required in order to provide the optimum level of public goods, such as new information. However, the difficulty with the Coase argument

when applied to new technology is that the legal costs to private firms of appropriating the returns on their R&D investments may be so high that only the government can provide certain types of information efficiently, for example, unskilled-labor-intensive technologies. Even "well-defined" legal rights do not guarantee a socially optimum level of appropriability.

The transactions and legal costs of establishing property rights for even sophisticated technologies are high. The irony is that private expenditures by individuals and firms to prevent the loss of appropriability are also public goods. The first firm in an industry may expend large sums to establish proprietary rights and establish legal precedents for property rights to complicated technologies used in the industry. Since subsequent innovators do not share in these investments but benefit from the appropriability protection they provide, they take a free ride on these legal investments. Thus, appropriability investments by MNCs may be low unless the industry is concentrated or some other consideration dictates that entrants pay "their share" of the appropriability costs.

There is ample evidence that the transactions costs of protecting new ideas are not trivial. This is true even for MNCs, which specialize in creating new technologies.[12] As a result, the company is forced to make costly variations in the patent on the same invention or process from country to country, making the descriptions wider or narrower, because of the local laws. Whenever a patent or licensing agreement is

11. Ronald H. Coase, "The Problem of Social Cost," *Journal of Law and Economics*, 3:1-44 (Oct. 1960).

12. W. M. Carley, "Multinational Firms Find Patent Battle Consume Time, Money," *Wall Street Journal*, 24 June 1974, pp. 1, 17.

found invalid in one country, this upsets licensing agreements the company has worked out in other countries.

There are several implications of the appropriability theory for MNCs. Next we examine how it explains why technology-creating firms, such as MNCs, are large. After that, we examine the light appropriability sheds on the effects of dismantling large MNCs. Finally, we examine its consistency with the stylized facts of foreign direct investment.

Appropriability and firm size

Appropriability is the first reason why firms that develop new products become large. Innovating firms expand to internalize the externality which new information creates, namely, the public-goods aspect of new information. A U.S. MNC whose technology is being copied in Western Europe will be better able and more inclined to stop the interloper if it has a subsidiary in Western Europe than if it exports the good from the United States or if it sells it through a marketing licensee in Western Europe. A European licensee selling only in Europe is less likely than an MNC to expend funds to stop a European interloper from exporting to, say, South America, since the licensee derives no benefit from this expenditure.

Second, there is a tendency for new products to be experience goods and for standardized products to be search goods. Optimum firm size is usually larger for domestic retailers of non-brand-name experience goods. By analogy, subsidiaries of multinational corporations are more likely than licensing arrangements. Third, sales of many high-technology products must be accompanied by sales of service information. The firm's optimum size is expanded because of service subsidiaries, for example, IBM's servicing of computers. Fourth, the number of products produced by information-creating firms is large because of economies of scale within each of the four types of information: development, production, marketing, and appropriability. Fifth, for new and differentiated products, the spread between the buyer and seller valuation of new information is higher than when the products are older and more standardized. This again suggests that market transactions costs are relatively higher earlier in a product's life cycle so that optimum firm size will fall through the industry cycle.[13]

These five reasons explain why new technologies are correlated with concentrated industry structures, why international trade in technology occurs within large MNCs rather than through licensing agreements, and why older industries are more competitive and less innovative. The correlation between concentration and R&D early in an industry's life and competition with less R&D late in each industry's life is now explained, although the direction of causation is not. The empirical evidence shows that firm size is negatively related to the average age of industries.

Dismantling large innovating firms

Given the interrelationships among the information needed for product creation, development, markets, appropriability, and

13. See Magee, "Technology and the Appropriability Theory of the Multinational Corporation."

related activities, the foregoing analysis suggests that dismantling large innovating multinationals might be very costly. The policy-maker must establish that there is an economic argument why the normal loss of appropriability and decline in optimum firm size through the technology cycle should be speeded up.

Appropriability and the stylized facts of foreign direct investment

The previous discussion provides a framework within which to interpret recent discussions of the multinational corporation, direct investment, and technology transfers. The framework here emphasizes that multinational corporations generate new products requiring large investments in four complementary types of information and careful calculations of the appropriability of each type. It is fruitful to treat technology like any other tangible good and to think of the international operations of multinational corporations as international trade in this commodity.[14] The revenue from trade in information is the present value of the monopoly profit streams permitted by international patent agreements and trade secrets. The price of the information is the monopoly element in the price of the new product. What implications follow from this approach?

Since international trade in information is analogous to international trade generally, both exporters and importers will play optimum tariff games.[15] Importers will tax it—to push down the price paid to exporters—and exporters will restrict its flow in an effort to raise its price; for example, the opposition of the U.S. government to General Electric's sale of jet engines to France in 1972. Technology importers should realize that if they try to lower the price they pay to foreigners for their purchases of information, this will increase their welfare but will reduce the quantity of technology imported below free trade levels—that is, reduce the "transfer of technology."

What is a "technology gap?" It exists in any situation in which a country is a net importer of a product, since less is produced than is desired domestically at world prices. Some regions have comparative advantages in creating information and others have comparative disadvantages. The theory of comparative advantage applies to trade in information just as it applies to steel, autos, and textiles: countries that do not have a comparative advantage in creating it should import it. Thus a technology "gap" should not be judged equivalent to a welfare distortion.

The phrase "transfer of technology" must be refined. First, the connotation of a costless gift should be discarded: all information transfers entail some cost. There are many ways in which information is transmitted: intrafirm transmission through the multinational corporations, market transfer through licensing, and government transfers through aid. Second, for the multinationals, we have already emphas-

14. G. K. Helleiner, "The Role of Multinational Corporations in the Less Developed Countries' Trade in Technology," World Development, 3:161-89 (Apr. 1975).

15. Carlos Rodriguez, "Trade in Technological Knowledge and the National Advantage," Journal of Political Economy, 83:121-35 (Feb. 1975).

ized that several types of information are created and transferable, and the type of information transferred should be specified. Third, the fact that existing information is a public good does not mean that speeding up its transmission is a welfare improvement. For example, a policy-imposed speed-up in the transfer of sophisticated production technology may cause its premature introduction into unskilled-labor-abundant LDCs.

Public versus private creation of new technology

Debate continues over whether research and development should be done by private firms or by the government. Proponents of government R&D point to the results of agricultural research sponsored by the U.S. Department of Agriculture. New seed varieties and other discoveries are disseminated widely to farmers at low cost. If the same results were privately controlled by patents, the price charged for the breakthroughs would be higher, reducing their dissemination and utilization and the social benefit from the research would be lower. Proponents of R&D by private firms—and its protection by patents or trade secrets—note that the market provides a superior allocation of R&D and speed of research effort from many new products. It is apparent that R&D will be better allocated among hairdryers, razor blades, and steel-belted radial tires by private profitability considerations than by government agencies. To summarize, it appears that private firms have an edge in deciding the products on which to spend R&D funds, while free government dissemination of the fruits of R&D is superior after breakthroughs are

made. A solution utilizing the strengths of each would be to let private firms do the R&D and have the government buy up the most successful patents and make them available to all interested producers, not just the innovator. The latter would compete the price down from near monopoly levels with patents. Despite its reasonableness, this compromise is fraught with difficulties: it is hard for the government—or anyone else—to know the value of new discoveries; the presence of a single seller—the innovator—and a single buyer—the government—generates nontrivial haggling costs; and government procurement is plagued with influence peddling and lobbying.[16] Sales of new military technologies to the U.S. Department of Defense illustrate both problems.

From Vernon's product cycle to the industry technology cycle

Vernon's product cycle suggested that the life of each product can be broken into three distinct stages: the new product, the maturing product, and the standardized product. He suggested that the locus of production would move from the originating developed country in the first stage to the LDCs in the third stage. Three considerations suggested why production would begin in the DCs for new products. On the demand side, high unit-labor costs generate demand for labor-saving investment goods and high incomes generate demand for sophisticated and differentiated new consumer products that save on household labor.

16. William Brock and Stephen P. Magee, "The Economics of Special-Interest Politics: The Case of the Tariff," *American Economic Review*, 68:246-50 (May 1978).

On the supply side, the research intensity of new products is high, and the relatively large endowments of skilled labor—scientists, engineers, and so on—dictate that DCs have a comparative advantage in creating new products. Finally, demand and supply interact, since rapid changes in new products require swift and frequent communication between producers and consumers.

In Vernon's stage II, production expands from the originating DC to other DCs as foreign markets grow, as other DC import barriers rise, as international transportation costs become a larger proportion of the product price, and as the production process becomes more standardized. In Vernon's stage III—the standardized product stage— production shifts to the LDCs since little interaction is needed between producers and consumers, small inputs of research and development are required, and as the production technologies become routinized through assembly lining so that more unskilled labor can be utilized in the production process.

How do we make the transition from Vernon's product cycle to an industry product cycle? First, evidence that industries may also go through cycles is suggested by data showing that industry patents follow an S-curve over long periods of time. Second, the appropriability theory suggests that industry structure itself is affected by these forces and varies systematically through the life cycle. Industry structure and R&D have a two-way interaction: concentrated industry structures encourage R&D—product creation, development, production and marketing; but also successful inventions encourage expansion of firm size as innovators attempt to

appropriate the returns on their R&D. The two-way causation suggests that young industries—those with new products—are concentrated, are associated with high R&D, and are more innovative while older ones are more competitive, spend less on R&D, and produce more standardized products.[17]

THE PRICE OF NEW HIGH-TECHNOLOGY PRODUCTS

Consider a multinational corporation that discovers a new product or a new technology that is embodied in a physical product. What price should it charge for this new product? If the innovating firm, such as IBM, charges a high price for the computer, it will experience high short-run profits but low long-run profits because of more rapid entry. Thus it gains short-run profits but gives away long-run profits. If it charges a low price in the short run, then it will have larger long-run profits but will earn low profits in the short run. It should be concerned about profits in both periods. The profit-maximizing firm will want to devise a pricing strategy that trades off short- and long-run profits appropriately. Such a strategy will be the one that maximizes the entire present discounted value of the monopoly profits on the new technology.

In a model more fully developed elsewhere, these elements and appropriability are integrated, producing the following propositions.[18]

1. The profit-maximizing price strategy for the innovating multinational is to set an initial price

17. Carley, "Multinational Firms Find Patent Battle Consume Time, Money."
18. See Magee, "Application of the Dynamic Limit Pricing Model."

slightly below the monopoly price and continuously cut this price through time until the long-run competitive price is reached.

2. An ironic result is that a legally imposed reduction in the long-run market power of an innovating multinational raises today's price of technology and reduces the short-run flow of technology to the developing countries. This occurs because the restriction reduces the value of future profit to the MNC. It attempts to make more profits in the short-run by raising today's price. At the higher price, however, fewer high-technology products will be purchased by the LDCs, and hence less technology is transferred.

3. Revisions in the Paris Convention that relax the enforcement of patent laws in favor of emulators would do the following. Since the innovating firm's profit stream is hurt by this more rapid entry, it becomes more concerned about its future profits. It will attempt to protect these profits by lowering today's price and all future prices in order to discourage the emulators. This results in an increase in short-run technology transfer and increases in the growth rate of technology transfer. However, there is no effect on long-run technology transfer by the policy of weakening patent enforcement, since the long-run price is unaffected.

4. There is a conflict between the level and the growth of technology exports. If the United States institutes a policy increasing the level of high-technology exports but policymakers evaluate it by watching the growth rate of exports, following the initial increase in the level, they will think—erroneously—that the policy failed. For example, subsidizing the cost of computer technology transfer will increase the number of computers exported, for example, from 1000 to 1100 per year, but the yearly growth rate may fall from 7 percent to 5 percent, starting from the new level of 1100. The policymaker should be aware of another important structural implication of the previous point. While the level of technology exports will be lower, the growth rate of technology exports will be higher, (1) to low-income countries for whom the innovating firm's costs of technology transfer are higher and to countries who have higher tariffs on technology imports, and (2) to high-risk countries to whom the innovating firm applies a higher discount rate.

POLICY QUESTIONS RAISED BY
THE DEVELOPING COUNTRIES

The previous section enumerated several propositions related to the behavior of a profit-maximizing/technology-creating multinational corporation. We did not address the trade-off between the current supply of technology and the future supply of technology. If the developing countries attempt to tax technology imports, they should consider this trade-off. It is discussed in the following section. The next section addresses the restriction in developing countries on foreign ownership in MNC subsidiaries in their countries. Multinational corporation limitations on R&D in developing country subsidiaries are discussed in the subsequent section, and the final section deals with excessive pricing of technology.

*Optimal taxation of
technology imports*

The conflict between the pricing of existing technologies and the creation of future technologies can be illustrated by a discussion of the related question of expropriation. It is clear that any country that expropriates a foreign-held firm increases its short-run welfare—since profits on the existing operation are transferred from foreigners to the domestic government—but reduces its future welfare—there is a reduction in future investment by foreigners in the country. The same is true of technology. Countries cutting the prices they pay for existing technology—increased restrictions on profits and repatriation, and so on—gain in the short run, but they reduce the future supply of new technology to the country. The LDCs as a whole face a difficult decision on technology import tax policy.

What criterion should be used? There is some technology tax rate that will maximize LDC welfare. However, this tax will be less than the standard optimum tariff in international trade theory. That tax ignores the future and is determined by increasing the tax rate until the marginal increment to home welfare because of that part of the tax paid by foreigners—the price paid to foreigners falls as the tax rises—just equals the marginal increase in consumer welfare loss caused by the distorted higher domestic price—the technical term for the latter is the "consumer's deadweight loss." However, with technology imports the future supply is affected so that we must change the criterion to apply to present discounted values rather than current flows: the tax should be increased until the marginal present value of the gain—the tax paid by foreigners—equals the present value of the marginal consumer's surplus distortion. Supply falls through time as the foreign supply curve shifts to the left.

Majority ownership of foreign affiliates

One of the most serious restrictions LDCs impose on multinational corporations attempting to set up subsidiaries in their countries, through either mergers or takeovers, is that foreign ownership not exceed 50 percent. The LDCs might benefit greatly if they permitted majority foreign ownership, at least 51 percent. This would still allow equity holders in the LDCs to capture nearly half of the profitability from the technology while overcoming the severe appropriability problem that is a key to much foreign direct investment in some high-technology industries. MNCs have a legitimate fear that if they do not control the operation, the other party to the agreement might "steal" the technology, sell it in third markets, and reduce the worldwide return on a given technology.

Limitations on R&D in LDC subsidiaries

Two restrictions on LDC research and development documented in a UN study[19] are the limitations by multinationals on R&D done in LDC subsidiaries or affiliates and in "grant-back" provisions for R&D done in LDCs—that is, that the host country must provide the parent with the results of subsidiary R&D on a unilateral and frequently unremunerated basis. This allows modifications and adaptations of sophisticated R&D to developing countries to revert to the parent.

19. United Nations Conference of Trade and Development, *An International Code of Conduct on Transfer of Technology* (New York: United Nations, 1975) TD/B/C. 6/AC. 1/2 Supp. 1/Rev. 1, pp. 34-35.

This is socially wasteful, since it prohibits LDC subsidiaries from transforming sophisticated ideas into forms more useful to the LDCs. However, it is explainable, since the appropriability of the latter is low. The parent realizes that the simpler technology may undercut its profits on the more sophisticated technology. There is no easy solution to this problem; anything permitting the substitution of less appropriable technology reduces the long-run supply of sophisticated technology.

Excessive pricing

A frequent complaint of technology importers is that the prices charged are "excessive." This could refer to attempts by technology exporters to extract the entire area under the technology import demand curve. If suppliers were completely successful in doing this, the quantity of technology transferred would equal that under free dissemination of technology. However, all of the economic benefit—technologists refer to this as the "economic surplus"—of the new technology is captured by the exporter rather than by the importer. The LDC obtains no welfare gain from the technology import, although he experiences no loss. One technique by which MNCs accomplish this is via a provision imposed on many technology importers in that the purchase of technologies be accompanied by an agreement to purchase—in some cases unrelated —raw materials, spare parts, intermediate products, and capital equipment from the technology supplier. A UN study showed that in four of the five importing countries, 66 percent or more of the contracts required tied purchase provisions.[20]

The "excessive pricing" problem is a normal consequence of the patent system: monopoly profits are the only reward of innovators, so that eliminating monopoly profits is an infeasible proposal if we expect private markets to create new technology. In fact, the case of the innovating firm capturing all of the benefits of a new technology cannot happen if there are active emulators. The model of optimal pricing by MNCs discussed earlier shows that an innovator will earn profits above production costs—otherwise it would have been foolish to extend the R&D funds—but these profits would be lower than those of a pure monopolist.

In the appropriability framework, innovators and emulators play important economic roles. Innovators reduce competition and charge high prices for new products, but they do innovate. Emulators steal technology and discourage innovation, but by cutting prices they make new technology available to us all. Private competition will provide for innovation and its dissemination properly only if the legal system and national policy balance appropriately the economic rights of innovators and emulators.

20. United Nations Conference on Trade and Development, *Major Issues Arising from the Transfer of Technology to Developing Countries* (New York: United Nations, 1975), TD/B/AC. 11/10/Rev. 2, p. 16.

Creation of Technology within Latin America

By SIMÓN TEITEL

ABSTRACT: The literature on technology transfer has traditionally assumed that the creation of new technical knowledge was restricted to the industrially advanced countries. Recent evidence seems to point in another direction since semiindustrialized countries are beginning to create new technologies, and even to sell technical knowledge, not only to other developing countries but also to industrialized countries. The available evidence based on enterprise and industry studies in Latin America indicates the existence of a fair amount of technological creativity, first in the form of adaptation of imported technologies to local conditions, followed by technological developments leading in a number of cases to the creation of new products and processes. Technological creativity arises in both competitive and noncompetitive industries, in large and small, public and private enterprises. While the existence of specific engineering and technical skills and of technically motivated entrepreneurs seems to have been crucial to the success of the technical changes taking place in Latin America, explanations of the factors accounting for these types of technological developments are not yet very robust, and a greater number of detailed case studies is required in order to acquire a fuller comprehension of the processes of technology transfer, adaptation, and creation in semiindustrialized countries.

Simón Teitel received his undergraduate education at the University of Buenos Aires, and earned a Master's degree in industrial engineering and a Ph.D. in economics from Columbia University, New York. Before joining the Inter-American Development Bank in 1968, he worked for the United Nations in New York and Vienna. In addition to his work as senior advisor in the Economic and Social Development Department of the Inter-American Development Bank, he is an adjunct professor of economics and business at the Catholic University of America. He is the author of a number of articles on industrialization, international trade, and technological development.

NOTE: I am grateful for comments received from S. Aquino and H. Schwartz. The viewpoints presented in this article are those of the author and do not purport to represent the official position of the Inter-American Development Bank.

What is knowledge? . . . It seems that there is something wrong with the ordinary use of the word "knowledge". It appears we don't know what it means, and that therefore, perhaps, we have no right to use it. We should reply: "There is no one exact usage of the word 'knowledge'; but we can make up several such usages, which will more or less agree with the ways the word is actually used."[1]

It had been taken, almost as a given, that an underdeveloped country does not create technology—that is, technical knowledge. Furthermore, the word "technology" generally conjures up visions of atoms, rockets, and computers. In this article we try to break away from some of the constraints imposed by these language standards. Our use of the word "technology" will convey a meaning that is less lofty and related to apparently trivial undertakings. By technology we mean the information, of a technical and organizational nature, required to manufacture industrial products, and by "technological change," all modifications made to such information. In terms of economic development, we will be dealing with a part of the range of countries—that is, semiindustrialized countries; and in terms of technological development, with only part of the spectrum—that pertaining to industrial technologies adapted and created in those countries. The experience reflected upon will sometimes be hard to differentiate from that of a more fully industrialized economy, the technology referred to being that of the best international practice. In other cases, the technology will be merely indicative of some adaptation to domestic conditions exhibiting limited local creation of new knowledge. In this respect, examination of the Latin American technology-creation experience appears to offer hope for new insights about the development process.

THE ISSUE

About 25 years after the United Nations' publication, *Processes and Problems of Industrialization in Under-developed Countries* in 1955, we can look today at the question of technology transfer in Latin America from a rather different viewpoint than prevailed at that time, when the main problem seemed to be how best to effect the transfer of capital and technology from the industrialized countries to the less-developed countries. In fact, after one generation, we are now in a position to consider the relative success of two policies undertaken in the region after World War II: industrialization propelled by import substitution protective policies, and transfer of technology resulting from direct foreign investment, international technical cooperation, and human capital formation programs. While the failures and achievements of the trade and industrialization policies have been widely discussed in the literature, this has not been the case with technological development, which is the theme we address in this article.

The process of technology transfer that took place in the region was closely connected with the process of industrialization.[2] Absorption of the technology that was

1. L. Wittgenstein, *The Blue and Brown Books* (New York: Harper & Row, 1958), pp. 26-27.

2. Simón Teitel, "Notes on the Transfer and Adaptation of Technology in Latin America, with Special Reference to Industrial Development in the 50's and 60's," in *L'Acquisition des Techniques par les Pays*

transferred demanded, as a prerequisite, know-how for the use of the technical information and its adaptation when necessary to local conditions. Moreover, the transfer was not entirely passive, with production leading to technical change and the development of new technologies. Technological development has not only led to the acquisition of technical capabilities to absorb imported technologies, but has also resulted in the local development of new products, processes, and an array of scientific and technical capabilities. This is shown not only in the production and export of locally developed models of various consumer and capital goods, but also in an incipient flow of foreign investment abroad,[3] and in the sale of licenses, consulting services, and turnkey plants to other countries.[4]

Having been repeatedly told that less-developed countries were precisely characterized by their inability to utilize advanced technologies because of their limited skills and educational base, the reader is entitled to ask where all this technological activity has come from. While the characterization of these countries is by and large correct, some Latin American countries already possess a stock of scientific and technical manpower[5] that allows them to undertake production of a variety of industrial goods and even to improve them in various ways. The economic development process has proven to be quite dynamic and the accumulation of capital, both physical and human— as well as the learning that took place—already has rendered such results that the division of labor acccurately reflecting present comparative advantage worldwide would be quite different from that prescribed 25 years ago. In a sense, we have relearned that much of comparative advantage is manmade and consequently subject to substantial change as a result of different priorities in the development policies and human resource development programs pursued by different countries.

THE EVIDENCE

The reader may still ask, What evidence do we have to justify our assertion about the existence of substantial technological creativity in the Latin American region? We may point to the following: (1) having acquired, as shown by casual observation, the technical competence to manufacture a variety of consumer and capital goods produced for domestic consumption in these countries; (2) having achieved in a number of cases quality and cost competitiveness in international markets as indicated by a growing flow of exports of manufactures embodying modern industrial tech-

Non-Initiateurs, ed. M.M. Daumas, Colloques Internationaux du Centre National de la Recherche Scientifique No. 538, (Paris, 1973), pp. 185-212.

3. C.F. Diaz Alejandro, "Foreign Direct Investment by Latin Americans," in Multinationals from Small Countries, ed. T. Agmon and C.P. Kindleberger, (Cambridge: MIT Press, 1977).

4. J. Katz and E. Ablin "From Infant Industry to Technology Exports: The Argentina Experience in the International Sale of Industrial Plants and Engineering Works," IDB/ECLA Regional Program of Studies in Science and Technology, Working Paper No. 14 (Oct. 1978); and C. Dahlman, "Technology Exports from Mexico." Mimeographed, The World Bank, Washington, DC, March 1981.

5. Simón Teitel, "Towards an Understanding of Technical Change in Semi-Industrialized Countries," Research Policy, 10: 127-47 (Apr. 1981).

nologies[6]; (3) the emerging ability to transmit technical knowledge to others as measured by the exports of disembodied technology in the form of licenses, technical assistance and consultancy, engineering services for infrastructure projects, turnkey plants, and so on.[7] Additional evidence of an indirect—and perhaps explanatory—nature refers to the availability and utilization of scientific and technical personnel in R&D activities in Latin America.[8]

To provide a better idea of the nature of the phenomenon being described, it is perhaps best to examine some cases of technological development in the region. First, we present three such cases, all belonging to the same industry, namely, steel-making, one each from Argentina, Brazil, and Mexico.[9] Then we review three other cases, all belonging to the same country but differing in organizational structure. While describing these cases we will try to illustrate the difficulties inherent in making broad generalizations or upholding stereotypes of

the kind: public versus private and foreign versus domestic ownership, product versus process innovation, small- versus large-scale firms, adaptation vis-à-vis frontier type of knowledge, and so on.

Case 1: Argentina

The information for this case comes from a study about the Rosario plant of a private steel-making enterprise in Argentina, which produced steel billets and rolled products.[10] Through successful, mostly local, engineering efforts, the capacity of this plant was stretched way beyond the stated or design capacity, achieving in this way a level of output and productivity that, although below international standards, represented a very substantial improvement. Technical change and experimentation also permitted increases in the productivity of inputs, adjustments in the product-mix, improvements in safety, and so on.

The economic environment of this plant consisted of an industry operating under substantial tariff protection, and the innovations introduced were possible because of the availability of a qualified team of engineers and technicians. It could also be argued that easy access to this alternative accounts for the plant not being scrapped earlier and was in part responsible for continued production of an expensive billet. From the viewpoint of the diffusion of acquired skills, while their technical improvements could not

6. For data on the volume of such exports and their proportion in total exports, see the World Bank, *World Development Report, 1980* (Washington, DC: the World Bank, August 1980). Also see OECD, *The Impact of Newly Industrializing Countries on Production and Trade in Manufactures* (Paris: OECD, 1979).

7. See Katz and Ablin; Dahlman; and Sercovich.

8. Simón Teitel, "Indicadores del Desarrollo Cientifico-Tecnologico en America Latina," (Washington, DC: Banco Interamericano de Desarrollo, Febrero 1981); and idem, "Indicators of Scientific and Technological Development—A Statistical and Econometric Approach," paper prepared for the Second Latin American Regional Meeting of the Econometric Society, Rio de Janeiro, 14-17 July 1981, Washington, DC: Inter-American Development Bank (May 1981).

9. The selection of the industry was dictated by the availability of comparable case studies.

10. P. Maxwell, "Learning and Technical Change in the Steel-plant of Acindar, S.A., in Rosario, Argentina," IDB/ECLA Regional Program of Studies in Science and Technology, Working Paper No. 4 (Buenos Aires, Argentina, Aug. 1977).

be directly reproduced elsewhere because similar plants are not constructed anymore, the experience acquired by the engineering team in this plant could be used in other plants of the same firm and eventually in the industry and economy.[11]

Case 2: Brazil

The information for this case comes from a study of R&D in a Brazilian public steel-making firm[12] and from a paper on exports of technology from Brazil.[13] It is an integrated steel-making mill located in Minas Gerais and producing flat rolled products. It was started with Japanese know-how and has estabished its own R&D center within the plant. This technological research center not only was instrumental in substantially improving the productivity of the plant's installations, stretching output and economizing in critical coke inputs, but the firm has also developed technical expertise in operating and designing steel-making facilities, which it is now actively selling to other firms, in Brazil and abroad. As in the Argentine case, its

success seems to have depended on the development of a highly qualified team of engineers and technicians supported by a technically oriented management. While the market was essentially noncompetitive, this firm, like other public sector firms in Brazil, was required to operate as efficiently as firms in the private sector. Thus one of the main differences between the two cases is that while the Brazilian firm is an efficient flat products producer by international standards, the Argentine firm was not.

Case 3: Mexico

The information for this case comes from a study about Mexico's technology exports.[14] This private firm developed a process to produce sponge-iron by direct reduction using natural gas as the reducing agent. Their need arose because they were originally producers of beer and needed steel to manufacture the crown covers for the beer bottles. The steel had to be soft and maleable, thus of very good quality, and since their requirements were limited in volume, they used electric furnaces of small capacity and scrap iron to produce the steel. Because at times they could not get a steady supply of adequate scrap for their furnaces, they undertook development of a process to produce sponge-iron by direct reduction, bypassing the very capital-demanding blast furnace route.

Their expertise was continuously improved under enlightened management, which included MIT engineering graduates, and their services began to be requested to help build sponge-iron plants in other countries and also to market

11. A similar phenomenon is noted by J. Katz, M. Gutkowski, M. Rodriguez, and G. Goity, "Productivity, Technology and Domestic Efforts in Research and Development—The Growth-Path of a Rayon Plant," IDB/ECLA Regional Program of Studies in Science and Technology, Working Paper No. 13 (Buenos Aires, Argentina, July 1978).

12. C. Dahlman and F. Valadares Fonseca, "From Technological Dependence to Technological Development: The Case of the Usiminas Steel Plant in Brazil," IDB/ECLA Regional Program of Studies in Science and Technology, Working Paper No. 21 (Buenos Aires, Argentina, Oct. 1978).

13. F.C. Sercovich, "Brazil as a Technology Exporter, with Sectoral Studies on the Steel and Alcohol Industries," (Washington, DC: Inter-American Development Bank, Apr. 1981).

14. See Dahlman.

and develop uses for the sponge-iron obtained through the direct reduction method. When the plants demanded from abroad increased in size, they developed modular units that could be duplicated in parallel to increase output as needed. Thus the limited scale with which they started did not become a permanent obstacle. A key factor in the development of their process and its ulterior improvement seems to have been the engineering background of the owners coupled with entrepreneurial capacity and limited aversion to risks, as shown by allowing experimentation at the plant level.

What conclusions could be derived from these cases: that Latin America is a star performer in the steel industry? Certainly not; although growing, steel output in these countries is still quite low in absolute and per capita terms. That we have creators of new processes at the technological frontier? In one case, that of the Mexican firm, yes, but not in the others. Can we say that it is because they are all public or all private enterprises? That is not the case either. What they seem to have in common is the existence of certain types of human resources able to perform the technical work, and also to organize it and to give it a chance. Conceivably, in other cases, the managerial response could have emphasized nontechnical solutions to similar problems.

Country cases

We consider next three other cases, all in one country, Argentina. The first is a subsidiary of a U.S. transnational corporation in the chemical field. This firm operated several facilities; the one we refer to was a rayon-producing plant.[15] This plant was started up with U.S. technology that was continuously modified through local R&D and engineering efforts. The work of the plant's engineering team resulted in new product varieties, permitting the use of different raw materials, product quality improvements, and increases in productivity. All the technological improvements originated in market demands, although the market structure in which the plant operated was initially characterized by a monopoly followed later on by a period of oligopolistic structure. The results of the R&D team were not limited to the rayon plant. Through their research and experimental work, they in fact trained engineers for the enterprise as well as for the parent company. They also helped in the development of small textile suppliers for dyeing, finishing, and so on, operations which were transferred out of the plant. Finally, they also exported their rayon-producing technology to other plants abroad. On the negative side, it is likely that as a result of their technical prowess they were able to keep the rayon plant functioning longer than warranted, causing the substitution of rayon by other synthetic fibers, like polyester and nylon, to be postponed beyond the point at which it made sense from the point of view of the national economy.

The second case is a small private firm, which initially produced car clutches for the replacement parts market, and subsequently, also for the original equipment market when the domestic production of cars was launched in the sixties.[16] It

15. See Katz et al.
16. Simón Teitel, "Tecnologia y empresa:

operated with technology acquired under licenses from U.S. and European firms for the large variety of clutches it had to manufacture given the relatively large number of car models produced for the local car market. While small, the firm was functionally organized with their own technical services including engineering and experimentation. Most of their technical work consisted of adaptations to local market conditions. These included the following: new product varieties, adaptations to different materials, adaptation to smaller runs and more models, and attempts at parts standardization. The market in which it operated was oligopolistic when it was only producing spare parts, and monopolistic when it became the sole supplier of new equipment for the local producers of cars. As a result of its technological activity, its engineers and technicians acquired training useful in other similar precision metalworking activities, in their own firm, as well as in the automobile industry and eventually the economy. It had also begun to export their technology as part of direct foreign investments in various Latin American countries. As a diseconomy, or negative effect, it could be argued that its satisfactory technical performance in adapting to different materials and to smaller runs—due to the large number of models—resulted in a costly clutch for locally produced automobiles.

The third case is also a subsidiary of a U.S. transnational enterprise, producing telecommunications equipment.[17] The source of its technology was Europe and the United States. Since with few exceptions, import prohibitions and "buy national" policies obliged it to produce with local raw materials and equipment, it undertook considerable work of adaptation, to domestic materials in particular. By improving manufacturing processes, plant layout, and assembly methods, it was able to increase productivity in several instances. Moreover, the technical information received from abroad was not always sufficient, and considerable on-site experimentation and trouble-shooting took place. Locally hired engineers and technicians had their own technical self-motivation to improve operations, although it did not really constitute a higher management objective, especially at first, when the plant enjoyed a monopoly and there was also only one buyer for its products, the government telecommunications agency, which used cost-plus contracts. As a result of the domestic engineering effort, engineers and technicians were trained in the plant in a variety of precision, electromechanical operations required for the manufacture of telephone equipment and that have wider applicability in industry. Later on, these skills were useful not only to the firm, but also for application in various other industries. They also had the effect, as in the case of the automobile industry, of contributing to the development of domestic suppliers, particularly of materials. Finally, the firm was later able to export telephone equipment to countries in Latin America and elsewhere, as well as locally produced tools and dies that were competitive with international quality

el caso de partes del automovil," mimeographed, Banco Interamericano de Desarrollo (Washington, DC, Julio 1978).

17. Simón Teitel, "Tecnologia y trasnacionales: el caso de los telefonos," mimeographed, Banco Interamericano de Desarrollo (Washington, DC, Agosto 1978).

and price standards. An adverse result of the process of technological adaptation in the context of such a high level of import protection was probably reduced product quality due, for example, to material constraints that led to lower electrical insulation standards.

Table 1 summarizes some of the characteristics of the six cases reviewed; three in the same industry and three in the same country. In Table 2 several "conventional wis-

dom" hypotheses about technological change and technology transfer are contrasted with the experience reviewed previously. More about this follows in the next section.

THE EXPLANATION

Technological change introduced in Latin American industry seems to be an involuntary byproduct of the manufacturing output, as well as the result of deliberate decisions

TABLE 1
CHARACTERISTICS OF TECHNOLOGICAL CHANGE IN
SEMIINDUSTRIALIZED COUNTRIES—SELECTED CASES

CASE/ COUNTRY	PRODUCT(S)	PROPERTY	SOURCE OF TECH- NOLOGY	NATURE OF TECHNICAL CHANGE
Steel 1 (Argentina)	Steel billets and rolled products	Private national firm	Largely local	-Stretching capacity output -Changing product mix -Increase productivity
Steel 2 (Brazil)	Steel flat rolled products	Public firm	Initially foreign	-Stretching capacity output -Changing raw materials and improving their productivity -Redesign of equipment and optimization to increase productivity
Steel 3 (Mexico)	Sponge-iron	Private national firm	Largely local	-New process for direct reduction
Argentina 1	Rayon	Subsidiary of U.S. TNE*	U.S. plans local R&D and engineering efforts	-New product varieties -Use of different raw materials -Quality improvements -Increased productivity
Argentina 2	Clutches	Private national firm	U.S. and Europe	Adaptation to: -New product varieties -Different materials -Smaller runs and more models -Standardization
Argentina 3	Telephone equipment	Subsidiary of U.S. TNE	Europe and U.S.	-Adaptation to different raw materials -Increase productivity (process improvement) -Increase productivity (layout and process normalization) -New recuperation process

(Continued)

TABLE 1 Continued

REASON FOR TECHNICAL CHANGE	MARKET STRUCTURE	EXTERNALITIES AND OTHER BENEFICIAL EFFECTS	DISECONOMIES AND OTHER NEGATIVE EFFECTS
-Market demand -Market demand -Malfunctions and hazards	Oligopoly	-Trained engineers for enterprise and eventually industry and/or economy	-Expensive billet -Delayed replacement of equipment
-Market demand -Market demand -Technical self motivation	Oligopoly	-Technology sales to industry -Patents -Technology exports	Not apparent
-Technical motivation in solving market demand problem	Oligopoly with product-differentiation	-Technology exports -Patents	Not apparent
-Market demand -Market demand -Market demand -Market demand	1st period monopoly (late fifties) 2nd period oligopoly	-Trained engineers for enterprise (other product plants) -Developed suppliers of textile operations (dying, finishing, and so on)	-Delayed substitution of rayon by polyester and nylon
-Market -Market -Market -Market	1st period oligopoly (spares) 2nd period monopoly (new equipment)	-Trained engineers and technicians for working activities eventually for industry and/or economy -Technology exports	Costly clutches similar metal
-Market -Technical self motivation	1st period monopoly 2nd period duopoly	-Trained engineers and technicians for precision electro-mechanical production, eventually for economy -Developed suppliers of materials and equipment -Exports of equipment tools and dies	Lower quality product

to set up laboratories and institutes of technological research, by private firms and in the public sector. Thus it is neither fully independent of expressly designed government policies nor primarily a consequence of such direct government intervention. Creation of technology in Latin America has probably been a result of the interaction of many variables. Some key ones are briefly examined next grouped under the following headings: (1) entrepreneurial and firm characteristics, (2) skilled human resource supply, (3) protection policies, and (4) direct promotion policies.

TABLE 2
SOME CONVENTIONAL HYPOTHESES ON TECHNOLOGICAL CHANGE
AND TECHNOLOGY TRANSFER

HYPOTHESIS	EXPERIENCE
1. Technical change is carried out to cut costs by increasing productivity	Mostly to adapt to demand: through product varieties, quality changes, increasing output, and to different materials, size, and so on.
2. Technical change is induced by changes in relative factor prices	Generally market or output induced
3. Scarcity of technical skills, especially to adapt and to improve processes and products, obtains in LICs.*	The opposite may be true in some cases leading to a socially "suboptimal" use of resources (see also 7).
4. Technologies used in LICs are transferred from abroad	There have been, not only numerous adaptations, but local production of technology and exports of technology from Latin American countries
5. TNEs† use only the latest capital intensive techniques and equipment	TNEs used second hand machinery and adapted factor use to local availabilities and prices.
6. TNEs only transfer "well-stabilized" mature technologies	-Those for high-speed spinning of rayon came immediately to Argentina (1950) -Those for manufacture of telephone receivers and the Pentaconta system still had many problems when transferred
7. Technologies used in LICs are not adapted to local conditions	A great deal of adaptation takes place. In fact, maybe too much. Perhaps, some of these resources should be allocated to producing altogether new product and process designs.

*Less-industrialized countries.
†Transnational Enterprise.

Entrepreneurial and firm characteristics

Ceteris paribus, entrepreneurial and firm characteristics may be the variable with the largest explanatory value of the technological activity observed in Latin America's industrial sector, or at any rate, it would seem so from the small number of detailed case studies available for review. That is, under similar conditions and stimuli, firms will differ in the nature of their responses. Part of the difference can be attributed to special factors or characteristics that bias the response toward technological change instead of other avenues. This is not to imply that any of these factors operated in a vacuum or that counteracting, or reinforcing effects, due to other variables were not present. It is also not easy to present a straightforward typology of firms and/or entrepreneurs according to whether they engage in technological research activities or not. The conventional distinctions as to public and private, foreign vis-à-vis domestic, small vis-à-vis large, competitive vis-à-vis noncompetitive, and so on, while useful at times, do not seem to have general validity as predictors of technological activity.

The empirical evidence being accumulated includes successful instances of technological developments in public[18] and private enterprises[19] in the same industry, albeit in different countries. While it is true that many times subsidiaries of foreign corporations are reluctant to establish R&D laboratories in developing countries,[20] it is clear, on the other hand, on the basis of available evidence that foreign firms are in some cases not only developing new products and processes locally, but are even engaging in domestic and foreign sales of the technologies they have developed.[21]

The small versus large distinction also leads to paradoxical results. There are a number of cases of small firms based on entrepreneurial personalities, with technical talent, who were able to develop new technologies, organize production using them, and even export this know-how to other countries. Such cases have been identified in Mexico: Alba, and the furfural process,[22] H y L in direct reduction[23]; in Brazil with the Italian family firms in the metalworking sector[24]; in Argentina with cases in agricultural machinery, consumer electronics, pharmaceutical laboratories, and so on. At the same time, there is no need to reiterate that some large firms are

not prone to engage in technological development activities of their own.

Analysis of the interaction of market structure and technical innovation yields similar results in developing countries as in industrialized ones. No valid across-the-board generalization can be easily sustained. Import substitution may have led to noncompetitive market structures and become a disincentive to innovation efforts of the type that concentrate on cost reduction and improvements in the productivity of inputs. On the other hand, in some cases, R&D activities become one of the firm's responses to changes in the economic environments. Lack of price competition may have resulted in competition based on product quality differences (see Argentina 1, Table 1), or in the need to stretch output under conditions of restricted availability of new machinery and relatively inexpensive engineering skills (see Steel 1, Table 1). Thus these two Argentine firms, although oligopolistic, engaged in various kinds of technological research activities. Moreover, even a domestic monopoly in a given product or process may be challenged in the international arena and lead a firm to engage in R&D activity to protect its international position and even attempt to export its technology (see Steel 3, Table 1).

To sum up, it is not easy to discern which characteristics make a firm more prone than another to engage in R&D and be successful, or better yet, to incorporate technological research activities as part of its standard response to changing conditions and long-run challenges. Many generalizations have been attempted, but they are hard to corroborate as having strong explanatory value given the nature of

18. Usiminas and CSN in Brazil; see Sercovich.

19. Hojalata y Lamina in Mexico; Acindar in Argentina; see Dahlman, Maxwell.

20. See Teitel, "Notes on the Transfer and Adaptation of Technology in Latin America."

21. Belgo Minera in Brazil—see Sercovich, Dupont in Mexico and Argentina—E. Aubert Montaño, "Exportaciones de tecnologia de la industria quimica privada de Mexico." Mimeographed, Banco Interamericano de Desarrollo y Banco Mundial, Febrero 1981, and Katz et al.

22. See Montaño.

23. See Dahlman.

24. H. Nogueira Da Cruz, "Industria Zaccaria e Sideroter," BID-CEPAL-PNUD Programa Regional de Estudios en Ciencia y Tecnologia, Relatorio Parcial Preliminar (San Pablo, Oct. 1979).

technological innovation observed in Latin American semi-industrialized countries.

Skilled human resource supply

Skilled human resource supply seems to have played a critical role in all or most cases, but it is also hard to categorize. The critical technological *factotum* can be an entrepreneur, a hired engineer, an engineering firm, or a R&D team assigned the resources necessary to research and to experiment, either in the plant or in a pilot installation.

In countries where an ample supply of engineers and technicians is available, there would be more use of their skills and, generally, more possibilities of technological change introduced into industrial processes. Availability is, of course, generally materialized in costs, and the relative costs of engineering skills vis-à-vis that of new machinery may determine the extent to which national responses may differ as to technological activity, even in the same industry.[25] The availability of "cheap" engineering skills may even lead to undesirable overengineering, and to delaying new investment and the introduction of new and more efficient technologies from the national economy viewpoint.[26]

One of the apparent lessons of the case studies reviewed is that while technological skills are indeed critical, their place in the firm is not always the same depending on its size and division of labor, nature of the technology used, inclination or reluctance to hire consultants, as well as on the type of technological change activities carried out. That is, there are improvements in efficiency that are centered in the application of industrial engineering techniques of wide, almost universal, applicability,[27] while in other cases a more "hardware" oriented approach is required or preferred.

We have also learned that there are in Latin America great discrepancies, as between countries, in the level of their stock of scientific and engineering personnel.[28] Also, while Latin American countries are, from the point of view of income per capita, degree of industrialization, and so on, at an intermediate level between the industrialized countries and other developing countries in Africa and Asia, in terms of their relative utilization of the stock of available personnel for R&D activities, as well as in their spending in R&D as a proportion of gross national product, the region does not perform significantly differently and may even lag behind these other developing areas.[29]

25. J. Fidel, J. Lucangeli, and P. Shepherd, "The Argentina Cigarette Industry: Technological Profile and Behavior," IDB/ECLA Regional Program of Studies in Science and Technology, Working Paper No. 7 (Buenos Aires, Argentina, Sept. 1978).

26. Teitel, "Towards an Understanding of Technical Change in Semi-Industrialized Countries."

27. A. Castaño, J. Katz, and F. Navajas, "Etapas historicas y conductas tecnologicas en una planta Argentina de maquinas herramienta," Programa BID/CEPAL/PNUD de Investigaciones sobre Desarrollo Cientifico y Tecnologico en America Latina, Monografia de Trabajo No. 38 (Buenos Aires, Argentina, Enero 1981).

28. See Teitel, "Indicadores del Desarrollo Cientifico-Technologico en America Latina."

29. Simón Teitel, "Indicators of Scientific and Technological Development for Less Industrialized Countries," *Technological Forecasting and Social Change*, 11(4):349-69 (Apr. 1978).

Protection policies

As stated previously, industrial protection by means of tariffs, quotas, "compre nacional," and so on may have the effect of inducing technological change that is required to adapt to local conditions and inputs in the import-competing industries. Excessive protection may have the effect of generating, or helping to generate, too much or too little technological research. The first possibility will arise when technical adaptations, scaling down or output stretching, are required to make production possible.[30] The second, because excessive protection by restricting competition may thus reduce the need to improve product quality and efficiency in production.

However, it is not altogether clear how local technological activity depends one way or the other on the type and level of protection afforded to industrial activities. While as noted before, technological research has been carried out under both circumstances—that is, with and without protection—the nature of the R&D activity does seem to be affected. That is, the R&D undertaken may tend to be more adaptive and less innovative in cases of high protection, and more cost-reducing in the cases of relatively open economies. However, normative conclusions cannot yet be reached. Protection of domestic industries may be required for a long period of time in order to complete a cycle of technology transfer, adaptation to local conditions, and learning to improve quality and productivity. Such technological activity may require protection, and, indeed, without it production itself may become unfeasible.[31] Stretching the point a notch further, those local R&D activities not directly a by-product of production may also require a degree of protection of their own until they can be exposed to the competition of technologies imported from abroad. However, such protection should be selective and limited in time, since the importation of technology from abroad is many times the first step in local technological developments.

Where export promotion policies are enacted, there is generally a strong presumption that to meet the stringent tests of international markets, quality will have to be improved and costs reduced, leading to R&D efforts targeted to achieve these objectives. Production for foreign markets may also require changes in design, adaptation to new specifications, use of special packaging, and so on. It is thus clear that neither export promotion nor import substitution will have a monopoly on R&D efforts, but that the nature of such activities may be different in each case. In many cases, exports result from spillovers of production originally destined to the local market. Moreover, the learning acquired while trouble-shooting production for the domestic market may also prove useful in the work of technical adaptation for exports.

Thus one of the lessons learned in recent years as to the effect of various industrial promotion policies upon technological change is that no obvious generalizations are valid.[32]

30. See Teitel, "On the Concept of Appropriate Technology."

31. L. E. Westphal, "Fostering Technological Mastery by Means of Selective Infant Industry Protection," in *Trade, Stabilization, Technology, and Equity*, ed. M. Syrquin and S. Teitel (New York: Academic Press, forthcoming).

32. Not at least until we have more evidence from detailed case studies, and are able to test some hypotheses developed from them.

One cannot argue that there has been an efficient export-oriented Asian model—which has favored improvements in productivity and technical change activities—and an import-substitution Latin American model, characterized by inefficiencies and little technical change activity. In the case of the Latin American countries, particularly the relatively more industrialized among them, protection may have been excessive or carried on for too long in some cases, but it has not deterred from technical change activities. It may have affected, however, the nature of these activities, although it has not even deterred from achieving a flow, still modest, but growing, of technology exports.[33]

The initial export orientation of the Asian semi-industrialized economies, while predicated on cost and quality competition in selected product lines, may have been undertaken from a not sufficiently broad industrial base—compared with the Latin American countries, or India, which favored import substitution strategies—and more critical yet, without the background support of a sufficiently large stock of experienced engineers and technicians to perform the R&D activities demanded for adaptation of technologies across a broad industrial spectrum. If this proves to be the case, while it is to be expected that the Latin American economies will have to streamline their industries and increase competition to achieve international efficiency, the Asian countries will have to widen their industrial base to achieve the learning, diffusion, and cross-fertilization effects required to move to higher

stages in their industrial and technological development. More important, as it may prove to be the real constraint, they will have to enlarge and deepen their human resources reservoir in the scientific and engineering fields required for such an expansion. This demands investments with long periods of gestation, and since much of the technological progress is to be attained as a joint product with production, substantial time for learning may also be required.

Direct promotion policies

Besides the indirect effect of general protection of production, specific measures to promote industrial research and technological development at the firm and industry levels have been undertaken in Latin America. While there are indications of substantial technological creativity in Latin America, as noted through causal empiricism, plant case studies, and analysis of technology exports, it is hard to relate these results to specific direct promotion policies and instruments. Direct experience, as well as comparative studies,[34] show the relative inefficiency of various explicit policy instruments to promote scientific and technological applications for development. These promotional activities have generally been characterized by attempts to provide science and technology with a visible institutional central focus, generally a National Council for Science and Technology, and with direct measures like tax holidays and other incentives to spend in R&D

33. See Katz and Ablin, Dahlman, Montano, and Sercovich.

34. F. Sagasti, *Science and Technology for Development: Main Comparative Report of the Science and Technology Policy Instruments Project* (Ottawa, Canada: International Development Research Centre, 1978).

installations and projects, risk-sharing programs, building of technological research institutes, fellowship programs, and so on. In some cases, even plans for science and technology were adopted.

The results of the planning efforts, although they were largely indicative in nature, have hardly been noticeable. The reasons for this are many. An important one is that the outputs of science and technology activities are hard to define clearly. Another is that the interrelations between scientific and technological activities are not well understood, and those between technological activities and production are also not clear. Furthermore, the availability of scientists and engineers, in the various specialties, depends on investment decision in education made years before and not clearly and directly connected, in general, to specific production or even research needs. All this makes planning for science and technology even more difficult and elusive than trying to do it in other areas of economic activity.

Undoubtedly, certain policy measures, both indirect and direct, may have had a powerful effect on the amount and nature of technological activity taking place in the Latin American countries, although here too it is hard to generalize. Conventional wisdom type of assertions that the programs in the public sector tend to be inefficient while those undertaken privately are not so, or that most technological institutes are white elephants, are not generally true. Cases of successful R&D undertakings have been detected, both in the public and private realms, as mentioned previously.

Some public institutes are well related to productive activities, particularly to industry; others are not. Financial mechanisms providing subsidies for R&D programs have not been very successful, but they have not been tried for long and, given the nature of research activities, this should not surprise anyone. The market failure metaphor in relation to underinvestment in R&D does not work in the same way in advanced as in developing countries[35]; further, more refined promotion efforts may thus be required.

One area in which instruments are still lacking is that of the provision of technological services for small enterprises. While attempts have been made through various mechanisms, they have not yet resulted in a transferable model. Financing institutions have a stake in this and some technological research institutes have made attempts to cover the gap, but matters of coordination and delivery of services remain as stumbling blocks.[36]

While explicit promotional instruments used in Latin America do not seem to have made much of a difference in terms of the R&D effort undertaken, there clearly is potential for refining the instruments and achieving at least similar efficiency in their application as in advanced countries.

35. Teitel, "Towards an Understanding of Technical Change in Semi-Industrialized Countries."

36. Inter-American Development Bank, *The Inter-American Development Bank and Science and Technology in Latin America,* Background Document for the United Nations Conference on Science and Technology for Development, Vienna, Austria, August 20-31, 1979 (Washington, DC, 1 June 1979).

Indian Technology Exports and Technological Development

By SANJAYA LALL

ABSTRACT: Exports of technology by developing countries can serve as an indicator of their technological development and its effect on changing their comparative advantage in international trade. This article concentrates on technology exports by India, a country that has built up a substantial industrial base but that has suffered from slow and erratic economic growth and has persisted in a policy of inward-looking industrialization. It argues that India is the leading exporter of industrial technology in the Third World, in terms of the range, diversity, and complexity of the technologies it sells, and traces this mainly to the strategy of the Indian government to foster technological "learning" in the capital goods sector of the economy. Thus, while its highly protectionist strategy has created various inefficiencies, it has also led to the creation of substantial technological capability in the country. Other developing countries have much to learn—both positive and negative—from the Indian experience.

Sanjaya Lall received his first degree in economics from Patna University, India, in 1960. After this, he graduated with first class honors from Oxford University, took a Master of Philosophy degree in economics, and joined the World Bank in 1965. In 1968 he returned to Oxford and is currently senior research officer at the Institute of Economics and Statistics there. His main interest lies in the multinational corporation, a subject on which he has published numerous books and articles.

NOTE: This article draws heavily on research being considered by the author for the World Bank on technology exports by India. The sponsorship of the bank is gratefully acknowledged, but the views expressed here are attributable only to the author.

T HE process of technical change in developing countries is one of the least understood subjects in the study of development. Despite the attention that such matters as international transfer of technology to the Third World, the role of transnational corporations, the choice of appropriate technology, and the building up of local technological capabilities have received in recent years, we still lack a proper appreciation of the fact that considerable technical progress is taking place in enterprises in developing countries and that such progress has significant implications for their future role in the restructuring of world trade and industry.[1]

Technical progress in developing countries is, however, different in its nature from what is normally understood by "technological change" or "innovation" in the advanced countries. In the latter, there is a general tendency to identify innovation with major, discrete advances in productive knowledge. Clearly, developing countries are not innovating in this sense of the word. A great deal of technical progress, however, takes the form of minor innovation, of changes in products and processes in their production and application in the diffusion phase of new technologies—it is such minor innovation that occurs in developing countries as they import and assimilate new technologies. The fact that it is described as minor should not lessen its economic importance; on the contrary, every student of economic history stresses the fundamental role of minor innovation in the development and commercialization of technology.

The fact that developing countries have the capability to undertake extensive minor innovations in the process of assimilating a broad spectrum of modern technologies does not, of course, rule out the prospect that they will continue, for the foreseeable future, to be dependent on the advanced countries for significant inputs of new technology. Developing countries over time will tend to specialize in technologies to which their capabilities are best suited—relatively small-scale, less expensive, lower-income-based products and processes—while the developed ones will specialize in technologies at the "higher" end of the scale.

By the very nature of the phenomenon, these matters are difficult to study empirically. The sort of detailed plant-level research required has yet to be undertaken in most developing countries.[2] One method of gaining some knowledge is to analyze *exports* of technology by developing countries; these consist of sales of turnkey industrial projects, consultancy services abroad, and direct investments or joint ventures by firms from developing countries. These do not reveal the whole range of technical progress under way in the exporting countries, but do provide examples of technical learning where the technology has been assimilated, reproduced, and brought up to standards of international competitiveness. This article proceeds as follows: the first main section briefly describes the extent of Indian technology exports, drawing upon empirical work I recently conducted for the World Bank. The second discusses the determinants of technological development underlying these exports, paying particular attention to the role of technology policy. The last section draws the main conclusions.

1. A full discussion of this theme is contained in Sanjaya Lall, *Developing Countries as Exporters of Technology* (London: Macmillan, forthcoming).

2. A major exception is the work of a team financed by the Inter-American Development Bank and the Economic Commission for Latin America directed by Jorge Katz.

INDIAN TECHNOLOGY EXPORTS

India has emerged in recent years as a major industrial power in terms of the value of its manufacturing production. It has, however, also suffered from a poor record of growth in income, industry, and manufactured imports. Much of its industry suffers from substantial excess capacity and remains inward-looking and highly protected. In some major areas of activity it is technologically far behind international frontiers.

The evils of excessive import-substituting industrialization are now well recognized. What is insufficiently appreciated is that a parallel policy, which need not necessarily accompany the protection of domestic production, of protecting domestic technological development may yield significant benefits that are not revealed by aggregate data on industrial or export growth. What we need, then, is to look at technology exports themselves. Most of the industrializing developing countries now undertake some form of technology exports.[3] Some of the larger ones undertake technology exports (TE) in all its forms, while the smaller ones specialize in particular forms of TE. Let us first review the evidence of Indian TE.

According to the patchy evidence at hand, India's TE at the end of 1979 stood as follows:[4]

It should be noted that different units of measurement have been used to quantify the different forms of TE and no simple method is available to aggregate them. Nevertheless, we can examine the sectoral distribution of these various stock and flow figures to gain some insight into the relative areas of revealed comparative advantage of Indian technology exporters. Table 1 sets out the percentage distribution of the three forms of TE by the main manufacturing and other activities.

In the most general terms, industrial project exports seem to cover a greater range of activities than other TE. Each form of TE does, however, have its own particular areas of specialization. For each, the two main activities account for over 60 percent of total foreign activity: power generation and distribution for turnkey, metallurgy and power generation for consultancy, and textiles and paper and pulp for direct investments. The only activity that occurs more than once is power generation. Otherwise the different modes of TE seem to have different advantages abroad, due to their market demand conditions, to their own peculiar nature, or to the technological strengths of Indian industry.

Industrial turnkey projects (completed and in hand)	Rupees (Rs.) 8540 million (m.) ($1.1 billion)
Consultancy Earnings Abroad: 1975-76	Rs. 40 m.
1976-77	Rs. 75 m.
1977-78	Rs. 95 m.
1978-79	Rs. 152 m.
Total	Rs. 362 m. ($46 m.)
Value of foreign equity (in 192 ventures in production and under implementation)	Rs. 795 m. ($100 m.)

3. See Lall, *Developing Countries as Exporters of Technology.*

4. Note again that civil construction projects are excluded. The value of such projects, completed and in hand, by the end of 1979 came to about Rs. 19 billion ($2.4 billion).

TABLE 1

INDIA: INDUSTRIAL DISTRIBUTION OF PROJECT,
CONSULTANCY, AND DIRECT-INVESTMENT TE
(in percentages)

	INDUSTRIAL PROJECT EXPORTS (CUMULATIVE, END 1979)	CONSULTANCY EXPORTS (1978-79)	DIRECT INVESTMENT (VALUE OF EQUITY, END 1979)
	TOTAL	TOTAL	TOTAL
Manufacturing			
Textiles, yarn	2.6	2.0	28.2
Sugar	5.3	4.9	3.6
Other food processing	0.3	0.1	10.2
Cement	10.9	0.5	—
Steel mills, other metals	6.5	24.5	1.2
Chemicals	0.7	10.7	6.4
Paper and Pulp	0.5	—	29.4
Simple metal products	0.2	—	6.0
Machinery, machine tools	5.6	—	3.8
Power generation	28.2	14.3	—
Power distribution	27.1	—	—
Transport equipment	0.8	—	3.3
Electronic, telecommunication	0.9	—	—
Other mfg.	0.7	6.1	2.4
Subtotal	90.5	63.1	94.5
Other			
Steel structures	4.6	—	—
Water treatment, sewage	4.9	—	—
Railways	—	7.2	—
Other	—	29.7	5.5
Subtotal	9.5	36.9	5.5
Grand total	100.0	100.0	100.0

Thus turnkey exports are particularly suited to large infrastructural projects where foreign equity investment is rarely permitted and the export of discrete items of equipment is not as remunerative as a large package. Here the strength of Bharat Heavy Electricals Limited (BHEL), a public sector firm, and Tatas in the manufacture and installation of power-generating equipment, and of Kamanis and Tatas in power transmission equipment, has enabled India to win significant foreign orders. It may have been possible for these enterprises to set up foreign affiliates to manufacture their products, rather than to participate in their user industries, but presumably scale economies, skill

shortages, and government policies in the host countries have rendered this infeasible for the time being.

Consultancy exports tend to be especially important in process industries because of the central significance of basic process design and the associated detailed engineering in plant construction. In engineering industries the use of consultants is much more restricted to specific problem solving, and the basic technology transfer functions tend to be undertaken by the equipment manufacturer and/or the user, as either a turnkey job or a direct investment. Thus the specialization of Indian consultants in metallurgy, power generation, and chemicals partly reflects the nature of the consulting industry. It also, naturally, reflects the accumulation of experience in a number of enterprises sponsored by the government or encouraged by it to participate in project design together with foreign consultants.

The pattern of direct investments reflects the technological and managerial strengths of Indian industry, to the extent that any of these has been developed sufficiently to yield a source of international "monopolistic advantage" to the enterprise concerned.[5] According to the product cycle type of reasoning, a developing country's overseas investment should be in the simplest and most widely diffused technologies, and the significance of textiles, food processing, and metal products in Indian investments accords with this. However, the existence of significant investments in fairly complex technologies like precision tools, pulp and paper, trucks, tractors, jeeps, and other automotive products—as well as projects like minicomputers in the pipeline—calls for a more cautious analysis.

The main weaknesses of Indian enterprises, used to protected seller markets, seem to be on the marketing side; this reinforces the impression that these enterprises' advantages lie in the technological and managerial spheres. Their technological advantage clearly does not arise from advanced research and development (R&D) capabilities in the sense that it does, say, for U.S. multinationals. It is, rather like early Japanese foreign investments, based on the ability to reproduce and commercialize a technology,

5. This point is analyzed at length in Sanjaya Lall, "The Export of Capital from Developing Countries: India," in *International Investment and Capital Movements*, eds. J. H. Dunning and John Black (London: Macmillan, 1982).

which may be fairly complex and large scale, in the peculiar conditions of less-developed countries (LDCs). We find examples where the techniques and products involved are intermediate in the sense of not competing directly with developed country enterprises; however, we also find examples where they are advanced and compete with identical products made by established multinationals, such as rayon, minicomputers, carbon black, and so on.

For all forms of TE, intermediate producer goods—power generation, power distribution, paper, steel, and chemicals—tend to dominate over consumer and capital goods as far as the end user is concerned. The production of consumer goods abroad is not, apart from fairly undifferentiated products like textiles and sugar, important because of the protected nature of the Indian home market and low average incomes. That of capital goods is generally not in demand in most smaller and industrially less-advanced host countries; where it is, much of it is in higher technologies than Indian firms can offer or can persuade buyers to accept. The major machinery producer overseas, Hindustan Machine Tools (HMT), a public sector firm, specializes its turnkey and direct investment activity on products that are below the capabilities and specialization of Western companies, and its own frontier R&D efforts are directed to products that are exported as such to developed countries, not in the form of TE to other developing ones.

Let us now quickly look at the destinations of Indian TE. Table 2 sets out the broad geographical distribution of the three forms of industrial TE as well as civil construction. It shows strikingly different patterns

TABLE 2
GEOGRAPHICAL DISTRIBUTION OF INDIAN TE
(in percentages)

	MIDDLE EAST	AFRICA	SOUTHEAST ASIA	OTHER (INCLUDING UNALLOCATED)	TOTAL
Industrial project exports	54.0	13.8	25.3	6.9	100
Civil construction	95.7	1.3	3.0	—	100
Consultancy	23.5	26.9	11.4	38.2	100
Direct investment	1.5	30.0	66.0	2.5	100

of TE: project exports, both civil and industrial, go mainly to the Middle Eastern oil countries, though industrial projects are considerably less concentrated than civil. Consultancy is fairly evenly spread, and the sale of such services as computer software leads to a large portion going to developed countries. Direct investments go mainly to Africa and Southeast Asia, though there are indications that a tiny share of the Middle East will rise in the future.

The reasons for such a pattern are obvious. The Middle East is the world's largest internationally open market for civil and industrial construction. Its high income, combined with a lack of local industrial skills and labor power, have led it to buy packages that most other regions would not. As the region grows industrially and achieves its desired infrastructural investments, the pattern of TE would change accordingly. Africa, on the other hand, has low incomes, low levels of industrialization, and low levels of skill. With the exception of Nigeria, it is not undertaking massive construction projects. Its needs for industrial technology are generally limited to simple consumer and intermediate products; again, Nigeria is an exception. Southeast Asia is a more advanced market in every way. There is very limited prospect for civil construction for obvious reasons; the advent of import-substituting industrialization rend-

ers direct investment the obvious form of selling technology.

So much for the review of Indian TE. It is clearly impossible, in the short space of this article, to compare Indian TE in detail with TE by other developing countries. Such a comparison has been attempted, albeit on rather scattered data, elsewhere.[6] What emerges from the comparison is that India seems to lead the Third World in the range, diversity, spread, and technological sophistication. Some countries have greater TE in particular forms: Hong Kong is a much larger exporter of direct investments and is particularly strong in textiles and consumer electronics. Brazil is more advanced in terms of petroleum and automotive technology, and Mexico in direct-reduction of steel technology; in the civil construction sector, of course, the Republic of Korea dominates the project export scene. But in general, India leads the others in the technological capabilities revealed in TE.

In view of the poor performance of the Indian economy this may cause some surprise; certainly, in the light of the existing view of import-substituting policies as inefficient and self-defeating, it calls for some explanation. Let us turn, therefore, to a consideration of the determinants of technological devel-

6. Ibid. and Lall, *Developing Countries*.

opment in a newly industrializing economy.

DETERMINANTS OF TECHNOLOGICAL DEVELOPMENT

The factors that determine the pace of technological development are a mixture of given economic conditions and official policies, with the two closely intertwined in most cases. Let us start with those that are least policy-determined and work up to those that are most so.

Size of market

The size of the domestic market—and, naturally, the length of industrial experience—is of crucial significance in affecting the nature of technological development because it determines the extent to which capital goods industries can be successfully established. It has been argued that every kind of production activity creates some kind of technological capability, from the smallest of village industries to the most complex of modern machine building. However, the potential for learning beyond small adaptations to imported capital equipment can be realized only if there is a domestic capital goods production base. Rosenberg, among a number of other scholars, has emphasized the crucial role of machinery—especially machine tools—manufacture in the generation and diffusion of technical change.[7] Clearly, the size of the home market, depending upon such conditions as size of country, per capita incomes, income distribution, and the like, will be a factor of prime importance. Exporting does, of course, offer the possibility of undertaking capital goods production

with large inherent scale economies even in small markets, but the complexity of basic capital goods design and engineering means that production for export will be either confined to simple labor-intensive operations within an international framework of production by multinational corporations (MNCs), as with electronics, or else be conducted by MNCs with designs entirely imported from abroad. While this may be conducive to economic growth for other reasons, as we have witnessed for a number of the newly industrializing countries, it will not give the economy a base for the export of the entire technology involved. For most complex technologies a necessary period of domestic production is required if the entire technology is to be mastered.

Skill availability

The importance of having sufficient numbers of technically trained personnel who can act as the receptacles of learning at the higher levels, and who can then transform problems of production and application into feasible solutions, that is, innovations, is so obvious that it hardly needs stressing. We need not dwell on this at greater length, but it takes us into broad fields of education, science, and training policy which are well known in their own right.

Promotion of local R&D

The explicit encouragement of technological work by a government can take two general forms: the creation of a research infrastructure not directly related to the production system and the encouragement of R&D within production enterprises themselves. The Indian government has gone into both, with

7. Nathan Rosenberg, *Perspectives on Technology* (Cambridge: Cambridge University Press, 1976), pp. 141-50.

TABLE 3
R&D EXPENDITURES AND EMPLOYMENT IN INDIA

	R&D EXPENDITURES (Rs. m.)	R&D AS PER- CENTAGE OF GNP	TOTAL SCIENTIFIC AND TECHNICAL EMPLOYMENT IN R&D
Year			
1958-59	290	0.23	20,724
1968-69	1310	0.44	73,634
1971-72	2140	0.54	103,767

SOURCE: National Committee on Science & Technology, An Approach to the Science and Technology Plan (New Delhi, 1973, Table 1).

special emphasis on the setting up of a large number—34—of national research laboratories under the Council of Scientific and Industrial Research (CSIR) responsible for developing a large variety of industrial and agricultural technologies, with special emphasis on atomic, space, and electronic technology, and on the commercialization of domestic innovations by the National Research Development Council (NRDC) and the Inventions Promotions Board. Table 3 shows total R&D expenditures in India in the public and private sectors up to 1972, and the number of scientific and technical personnel employed in formal R&D activity.

Table 3 shows that the formal R&D effort has grown rapidly in recent years, increasing over sevenfold in terms of expenditure—in current rupees—and fivefold in terms of the technical personnel employed, and more than doubling its share of gross national product (GNP). Later estimates show R&D to be reaching 0.9 percent of GNP by 1978-1979. While the effort is small in relation to developed countries, in absolute terms it is quite large and not unimpressive for a developing country.

Formal R&D figures may, however, be misleading for two opposing reasons. First, they may overstate the amount of technological work that is applied to production. A large part of the money may be spent on scientific work that is unrelated to the production system, given the gaps that exist between the scientific and industrial establishments of developing countries. Second, they do not take account of technological work undertaken outside formal R&D institutions within the production enterprises in the normal course of their investment and manufacturing activity. As we have argued, and as has been documented in detail in Argentina, this is the chief source of technical progress in developing countries.

As for the promotion of R&D within manufacturing industry, in-house R&D has been encouraged by means of various fiscal and other incentives: highly accelerated depreciation allowances for research equipment; tax allowances of 120 percent of the current expenditures on R&D; privileged access to foreign exchange for the import of equipment, samples, and journals; and special awards for innovation. The tax incentives have been particularly successful in encouraging firms to set up separate R&D units or to seek official recognition for units already in operation. A few large firms—like the Tata Engineering and Locomotive Company (TELCO), though other Tata firms have recognized R&D units—have not sought recognition for their R&D, perhaps because of secrecy or a desire to avoid bureaucracy; however, almost all major firms now have recognized research facilities in both public and private sectors. By end of 1979 there were some 550 recognized units in the private sector and over 60 in the public sector. The number of recognized units was rising rapidly, and in 1977-79 over 100 units per annum were accorded this status.

The rapid growth of in-house R&D by manufacturing units, especially in the private sector, has been striking. However, industrial R&D comprises under 10 percent of total science and technology expenditures in the country, as compared, say, with 75 percent in the United States and 65 percent in the United Kingdom and Japan. The bulk of it is accounted for by the nonmanufacturing research establishments run by the CSIR, and here the returns in terms of commercial application in industry have been extremely poor. Thus the sheer quantity of resources, financial and human, invested in formal R&D activity in India cannot explain the buildup of its technological capabilities.

Protection of local learning

Factors such as large markets, experience, skilled manpower, and formal science and technology programs usually accompany technological learning. They are, however, necessary but not sufficient for such learning. A certain amount of elementary learning, that is, shop-floor productivity-raising improvements, is inherent to the manufacturing process and proceeds almost automatically regardless of who owns the facilities and what government policies are. Learning beyond this level to modify existing processes substantially to redesign the product or reproduce the facilities elsewhere, requires a more conscious, directed effort on the part of the enterprise concerned—either the final product firm or the capital goods manufacturer.

More importantly, advanced learning involves the undertaking of a new and inherently risky activity of technological innovation. Even if the technology is not new and is well known in developed countries, a developing country enterprise discovering a new product/process or redesigning capital equipment will face certain costs and uncertainties. Such costs will be met only if some of the uncertainties and risks are reduced by official intervention.

There are two main avenues for importing ready-made technology from abroad: direct investment by foreign enterprises and the licensing of local enterprises, and intermediate positions such as joint ventures. The direct entry of wholly owned MNC affiliates probably provides the most powerful means of importing a continuous stream of ready-made techniques and products based on the frontiers of technological work abroad. Licensing provides a channel of importing particular technologies for specified periods and is much more limited. Joint ventures lie between the two, the strength of links with foreign technology depending on the nature of participation, the complexity of the technology, and the relative capabilities of the two parties.

The import of technology in these forms can be said to stifle the growth of local technological capability if two conditions are met: first, if the technology is such that local efforts can lead to an internationally competitive and/or socially desirable technology and, second, if the technology importer does not himself undertake the investments locally to build up this capability. It may happen, for instance, that a multinational affiliate has to undertake local R&D work to adapt its processes to local climate, scale, or raw materials or to modify its product to local conditions, culture, or official regulations. In this case the affiliate may well generate as much technological activity as a local firm, cut off from access to foreign technology, would have done. Even more, it may well generate this activity more efficiently, with more resources and with better scientific backup from abroad.

It may, however, happen that the multinational affiliate—or joint venture or licensee—remains passively dependent on its foreign partner for all improvements to technology, or for all technological work beyond the level of minor adaptation or detailed engineering. In this case, a viable technical capability to assimilate, improve, and export the entire technology will not be built up in the developing country: the particular enterprise concerned will not find it economical to undertake the cost and risk of reproducing technological work already done—and proved—abroad, and its local rivals will not be able to compete on open markets if they try to build up their own independent technology. Any firm that is to remain competitive from the start would have to resort to foreign technology unless the industry was such that foreign technologies were totally irrelevant.

It should be noted that the protection of domestic production by tariffs against imports may go closely together with the protection of domestic learning, but it may not. Thus an import substitution regime may do little to build up an independent technological capability; on the other hand, an export-oriented regime may be strongly protective of local learning. The successes and failures of broad industrialization strategies such as import substitution or export promotion do not necessarily reflect their successes or failures in the sphere of technological progress. Japan provides an excellent example of successful export-promotion and high growth based on very strict protection of domestic learning.

Returning to the case of developing countries, it is argued here that what is broadly termed the "protection of domestic learning" is the crucial factor explaining India's apparent lead in indigenous technological capabilities. Of the newly industrializing countries, India is the only one that has consistently and over a long period pursued a policy of self-reliance in technology. It has placed close restrictions on MNC entry and on foreign licensing, and has forced local enterprises to invest in basic 'design capability. There is no doubt that this policy has had several costs: India is lagging well behind its competitors in several fast-moving technologies; several sectors have not succeeded in fully absorbing their technology; resources have been wasted in unnecessary effort and mistakes; and many local adaptations may not be internationally efficient. Granting all this, the policy does seem to have generated considerable learning in a broad range of activities and, as the preceding section of this article showed, it has led to the export of certain technologies where India has developed a competitive advantage.

The role of protection of learning also shows up clearly in some other developing countries that have TE. Brazil and Mexico, for instance, export mainly petrochemical and steel technology where the state has intervened to protect local enterprises, by setting up government-owned enterprises—Petroleos Mexicanos (PEMEX), and PETRO-BRAS—or by subsidizing private local enterprises. Argentinian enterprises have exported technologies mainly in relatively simple industries—food processing and metal products—where copying is easy and the large multinationals are not particularly active. Korean, Taiwanese, and Hong Kong enterprises similarly have exported the rather simple technologies where there is natural protection given by the wide diffusion of know-how. And exports of civil construction know-how have been fostered by the natural protection given by the nature of the activity and official policy. Korean enterprises have, however, advanced somewhat further because of the close links they have fostered with their government, their relatively large size and dominant positions in local production and exports, and the heavy official promotion of all forms of export activity.

CONCLUSION

Technological development is a complex phenomenon, and little is known about its "nuts and bolts" in developing countries. The arguments advanced here are based on rather scanty evidence and speculation and so must be treated with due caution. Nevertheless, the Indian case suggests certain lessons for other developing countries.

First, the size of the country must be a crucial factor in determining the pattern of industry that can be efficiently established in it. Small countries clearly must not try to emulate the patterns of heavy industrialization followed by large ones. The example of India, in particular, is not recommended, because it has gone too far and too fast in its efforts to build up a comprehensive industrial base. Moreover, many industries where it is now internationally viable cannot be launched by countries with smaller markets. This necessarily limits the scope of technological progress in small countries, but such are the dictates of technological influences on efficient scales of production.

Second, India has created a large pool of skilled manpower that has served as a receptacle for technological learning. It need not be stressed that a country's educational structure must be geared to producing the right numbers of appropriately trained personnel to undertake the task of technological progress.

Third, officially sponsored science and technology programs based in formal institutions divorced from production seem to be ineffective in promoting technological progress in industry. The bulk of effort must therefore be directed to stimulating technological activity in productive enterprises.

Fourth, the main message of this article is that, given all the other requirements of technological progress, the enlargement of indigenous capability in basic design and development work in industry necessarily requires some protection of local learning. This protection has positive as well as negative aspects: the former includes subsidization of R&D activity, protection of the market, and the like, while the latter includes a forced reduction of dependence on imported ready-made technologies and a constriction of the activities of foreign companies. It must be emphasized that there are large potential costs in following these policies, though successful cases of technological protection like Japan lead us to believe that a properly managed policy can yield tremendous benefits.

Finally, the protection of domestic learning must not be a wholesale or continuous policy. There are large areas of industry where local enterprises cannot master the requisite technology or, having mastered it, cannot keep pace with its development. In this case there must be a continuous inflow of technology from abroad—Japan is again an excellent example—complemented by local efforts to absorb and reproduce it. The real art of technology policy thus lies in identifying a country's dynamic comparative advantage in the absorption and generation of technology, and India's experience shows that the acquisition of this art may well be a painful and expensive process.

Technological Activities of Less-Developed-Country-Based Multinationals

By DONALD J. LECRAW

ABSTRACT: In recent years firms in developing countries have begun to undertake direct investments abroad. Their competitive advantage over other multinational enterprises, locally owned firms, and imports lay in their possession of proprietary technology that cannot be transferred via the market developed in response to factor market and demand conditions in their home country. This technology was appropriate to the host country environment: labor-intensive, flexible between products and processes, operated at high rates of capacity utilization, and intensive in the use of locally produced inputs. These investments have typically been motivated by rising labor costs at home and threats to export markets both in the host country and in third country markets.

Donald J. Lecraw is associate professor of economics and international trade, Centre for International Business Studies, School of Business Administration, The University of Western Ontario. He received a B.S. from Stanford University, an M.B.A. from Harvard Business School, and a Ph.D. in business economics from Harvard University. Professor Lecraw has served as chief economist, the Royal Commission on Corporate Concentration, and consultant to the Economic Council of Canada, the Institute for Research on Public Policy and the United Nations Centre on Transnational Corporations. He has published books and articles on industrial organization, choice of technology, economic development, and international business.

NOTE: Research for this article was partly funded by the Centre for International Business Studies.

TECHNOLOGY can be trans-
ferred between countries in
many ways: exports of final products, industrial inputs, and capital
equipment, turnkey projects,
governmental technical aid and services, technical communication,
training programs, consulting services, management contracts,
licensing, and foreign direct investment (FDI). This article focuses on
the last of these mechanisms of technology transfer—foreign direct
investment—particularly foreign
direct investment undertaken by
firms based in developing countries.
Three issues are addressed: the
types of technology developed by
these multinational enterprises
from developing countries (MDCs);
the mechanisms by which they
transferred their technology to the
host country; and the impact FDI
and technology transfer by MDCs
has had on both the home and host
country.[1] These three topics will be

analyzed within the context of a theory that explains the competitive
advantage of MDCs that enables
them to operate abroad in competition with multinational enterprises
based in developed countries
(MNEs) and locally owned firms in
the host country and the motivations
of MDCs for undertaking FDI in the
face of this competition.
    Before embarking on this task,
two points of caution should be
raised. First, except for a very few
host and home countries, the aggregate statistics on FDI on a worldwide basis are incomplete,
inaccurate, noncomparable, and
biased, usually downward. This
data problem is often severe for FDI
by MDCs. Conclusions based on
aggregate data on FDI can only be
viewed as highly tentative.[2] Second,
although the statistics are far from
clear, it can be safely concluded that
FDI by MNCs has been small, both
in absolute terms and in comparison
to FDI by MNEs based in developed
countries and to trade between developing countries. For example,
FDI by MDCs based in Latin America as a percent of total FDI in Latin
America ranged from 6.8 percent in
Guatemala to .22 percent in Mexico
and for eight countries was 1.2 percent of total FDI.[3] Even for major
source countries of FDI by MDCs,
such as Argentina, India, Hong Kong,

1. As yet, there is no commonly accepted
term for firms in developing countries that
have undertaken foreign direct investments.
Several authors have differentiated multinational enterprises based in developing coun-
tries (MDCs) from multinational enterprises
based in developed countries—referred to as
MNCs, MNEs, or TNCs, depending on the
author—for example, as follows: Sanjaya
Lall, "Exports of Manufactures by Newly
Industrializing Countries: A Survey of
Recent Trends," *Economic and Political
Weekly*, 6 and 13 December 1980; Peter
O'Brien, "Third World Industrial Enter-
prises: Export of Technology and Invest-
ment," *Economic and Political Weekly*,
Special Number (October 1980); and Donald
Lecraw, "The Internationalization of Firms
from LDCs: Evidence from the ASEAN
Region," in *Multinationals from Developing
Countries*, eds. K. Kumar and M. McLeod
(Lexington, MA: D.C. Heath, 1981). In this
article, MDC will refer to a firm based in a
developing country that has made a direct
investment in one or more foreign countries;
MNE will refer to a multinational enterprise
based in an industrialized—developed—
country.

2. Professor Louis T. Wells, Jr., at Harvard Business School, has assembled a data
bank on over 1000 FDIs made by MDCs. This
data bank has proved to be extremely valuable, although it is far from complete in cover-
age of all the MDCs or in the scope of the data
on the firms in the data bank.
    3. The data comes from P. O'Brien and J.
Monkiewicz, "Technology Exports from
Developing Countries: The Cases of Argentina and Portugal." Mimeographed. Geneva:
UNIDO/IS 218, March 1981, Table 7, p. 48.
The eight host countries were Argentina,
Brazil, Colombia, Ecuador, Guatemala, Mexico, Peru, and Venezuela.

and Singapore, outward FDI was typically less than one percent of their trade with other developing countries.[4] Despite its small size, FDI by MDCs is an increasingly important mechanism by which technology that has been developed in one developing country is transferred to other countries. It is particularly important, since, as described subsequently, this technology is often quite different in type and condition and means of transfer from that associated with MNEs based in developed countries.

COMPETITIVE ADVANTAGE OF MDCs

The size and global reach of MNEs from developed countries have often obscured a fundamental and simple fact of international business: if everything is equal, operating abroad either through exports, licensing, or direct investment is more difficult, time-consuming, risky, and costly than operating at home. Put another way, in order to operate abroad, a firm must have a powerful competitive advantage over locally owned firms in host country markets in order to overcome the cost disadvantages inherent in business activities outside its home country: tariff and nontariff barriers to trade, costs of coordination and information gathering and processing, government restrictions on trade, FDI, and licensing, and so on.

Until recently, most MNEs have typically overcome these cost disadvantages and been able to operate internationally through possession of proprietary process and product technology, marketing and management skills, and access to relatively cheap capital and to markets in their home country. The MNEs' proprietary "core skills" were often developed in response to demand in their home markets for income-elastic final products and large-scale, capital-intensive, often high-technology production equipment and facilities that substituted technology and capital for labor.[5] Panglaykim has identified two other strengths of Japanese MNEs that allowed them to operate internationally despite their initial lack of capital and technology: close business-government relations and cooperation and well-developed trading skills.[6]

The economic conditions that fostered the development of these MNEs, however, have not been present in most developing countries, where the market size has often been small, income low, labor in plentiful supply, and the level of technological and marketing sophistication and training below that in developed countries. Developing

4. This comparison of trade between developing countries and the flow of FDI by MDCs is somewhat misleading. Once the FDI is in place, it will produce a continuing flow of products and services. A more useful measure for purposes of comparison might be the stock of FDI by MDCs. This figure is not available, however, for most host and home countries.

5. This description of the competitive advantage of MNEs was particularly applicable to American and European MNEs in the manufacturing sector through the mid-1970s. More recently, it has lost some of its force; see the following: I.H. Geddy, "The Demise of the Product Cycle Model in International Business Theory," *Columbia Journal of World Business* (Spring 1978); and Raymond Vernon, "The Product Cycle Hypothesis in a new International Economic Environment," *Oxford Bulletin of Economics Statistics* (Nov. 1979).

6. J. Panglaykim, *Emerging Enterprises in The Asia-Pacific Region* (Jakarta: Centre for Strategic and International Studies, 1979).

countries have typically been net importers of product and process technology; their firms have not produced and marketed branded, high-quality products; capital was relatively expensive and there have often been restrictions on export of capital. Firms in developing countries have typically exported raw and semiprocessed natural resources and agricultural products and, if they have exported manufactured products at all, these products were low-quality, labor-intensive, undifferentiated products that competed on the basis of price, in sharp contrast to the products that have formed the basis for FDI by MNEs from developed countries.[7] What, then, is the basis of the competitive advantage of MDCs that has allowed them to succeed as direct investors in the competitive and often hostile markets in host countries abroad in competition with locally owned firms and MNEs?[8] Although the data are far from complete, some tentative answers can be offered.

At least initially, most firms in the modern manufacturing sector in developing countries have imported their product and process technol-

ogy directly from developed countries via FDI, imports of machinery, licensing, and so on.[9] In fact, there often seems to have been a direct relationship between the speed of growth of a country's modern industrial sector and its ability to absorb technology from abroad. Development of technology in developing countries has typically followed a path of importing technology "as is" from developed countries, to "learning by doing" and "learning by adapting" using that technology, to "learning by design" and "learning by improved design," to "learning by setting up complete production systems" and "learning by innovation," and finally to "learning by basic research and development."[10] As the experience and capabilities of firms in developing economies has increased over time, they have moved up this chain in the absorption, innovation, and development of technology both with individual products and processes and in their overall technological capabilities. The technological capabilities of firms have also increased as the economies of their home countries have developed and the level of skills, training, and education of the work force has increased. The capabilities of firms in developing economies to absorb, modify, and develop technology has been directly related

7. This generalization is somewhat overdrawn in the case of several developing countries with more advanced manufacturing sectors, as pointed out by Sanjaya Lall, "Developing Countries as Exporters of Technology," *Research Policy*, 9 (Jan. 1980). Firms in India, Argentina, Korea, and Singapore, to name a few, have exported quite sophisticated manufactured products and have bid successfully against MNEs for turnkey projects.

8. MDCs in the Association of Southeast Asian Nations (ASEAN) region not only were successful in their FDI, but their profits were higher than those of other MNEs and of locally owned firms, as shown in the following: Donald Lecraw, "Foreign Direct Investment by Firms from Less Developed Countries," *Oxford Economic Papers* (Nov. 1977); and idem., "Internationalization of Firms from LDCs."

9. The mix of inward FDI, imports, and licensing has varied significantly between countries; see Lall, "Exports of Manufactures."

10. This classification of the absorption and development of technology in developing countries is taken from Lall, "Developing Countries as Exporters," pp. 39-40. Elsewhere in this volume Lall describes the development of technology in much greater detail. For a similar classification, see also Linsu Kim, "Stages of Development of Industrial Technology in a Developing Country," *Research Policy* (July 1980).

to the level and depth of the industrial development of these home economies.

Exceptions to this generalization have occurred in some countries that have followed a strategy of industrialization based on "export platforms" and a high degree of FDI by MNEs that utilized locally available, low-wage, unskilled labor. This type of industrialization has often failed to foster the growth of the technological or manufacturing capabilities of either subsidiaries of MNEs or locally owned firms in the modern industrial sector, although it has often led to a rapid increase in industrial output and employment.

The direction of the imitation, modification, and adaptation of imported technology and the development of new technology by firms in developing countries has been in response to product and factor market conditions that have prevailed in their local markets. Typically, demand in these markets has been quite limited by both the size of the market and the oligopolistic nature of the competition that has prevailed between the few firms in each industry. To meet these conditions, some firms have adapted imported technology so that it could operate efficiently at a smaller scale than that for which it was originally designed in developed countries. This down-scaling in the size of equipment and machinery and reduction in the speed at which it was operated has been accompanied by three related modifications, all in response to factor and product market conditions in the home market.

First, production technology was modified so that it was more flexible between different products. Instead of dedicating a piece of equipment or production line to the high-speed, continuous production of a single product, production equipment and processes were modified or designed so that several products could be produced using the same equipment with low downtime and changeover costs.

In other instances, both production processes and products were modified so that the same product could be produced by different processes and equipment. Increased flexibility in production technology was also often necessitated by rapid changes in demand and supply conditions in developing countries. Efficient operations under changing demand conditions required both more flexibility between products and faster adjustment of capacity utilization than was typical in developed countries.

Firms in developing countries have often had difficulty obtaining the inputs either locally or on the international market that were appropriate for their imported technology due to lack of domestic suppliers capable of producing the inputs—at least in the quantities and qualities needed—and to import and exchange restrictions. These input supply constraints have forced firms to modify product and production technology so that they could utilize inputs that were available locally or to develop their own in-house capabilities to produce these inputs using raw materials that were available locally. Moreover, due to periodic supply shortages, firms have often had to modify production technology and processes to increase their flexibility between alternative inputs.

These modifications of imported technology and innovations of new technology in scale and flexibility between products, processes, and inputs have tended to make the production technology employed by firms in developing countries more labor-intensive than that employed by MNEs either in their home countries or abroad. In addition, the low-cost, plentiful supply of labor in most developing countries has induced firms to substitute labor for capital in the production process in order to reduce costs.[11]

On the product side, the generally low levels of income in developing countries have led to demand for low-priced, generally low-quality products with low marketing and technology intensities.[12] This demand has brought forth product modifications and innovations by some firms in developing countries. In turn, this product modification has both allowed and motivated firms to modify their production technology. In order to produce these "appropriate" products at lower cost, firms have often modified their production technology, usually in a more labor-intensive direction, since quality and uniformity were not as important for these products as was low cost.[13]

In summary, conditions in the factor and product markets in developing countries have motivated some firms in those countries, particularly locally owned firms, to undertake modifications and innovations of product and production technology to respond to the conditions that prevailed in their home markets. These firms have developed their own unique core skills in process and product technology and marketing and management that have given them a competitive advantage in producing products and using technology that were appropriate for market conditions in their home countries, and, as described subsequently, in other similar markets abroad.[14]

MOTIVATIONS FOR FDI BY MDCs

Even when firms based in developing countries have developed a competitive advantage that might enable them to succeed in markets in some countries abroad, the question remains of why they would service these markets through FDI rather than by some other type of

11. The conditions under which firms in developing countries were actually motivated to develop and use cost-efficient, labor-intensive technologies are analyzed in Donald Lecraw, "Choice of Technology in Low-Wage Countries: A Non-neoclassical Approach," *Quarterly Journal of Economics* (Nov. 1979).

12. The often highly skewed nature of the income distribution in many developing countries has often led to a demand for the same high-quality, branded, technology-intensive products that were available in developed countries. This demand has typically been serviced by imports or by local firms—either subsidiaries of MNEs or locally owned firms—that produced these products without any significant modification in production or product technology. See the following: Lecraw, "Choice of Technology"; and Louis Wells, Jr., and Pankaj Ghemawat, "The Generation of Industrial Technology by the Less Developed Countries." Mimeographed, Harvard Business School, August 1980.

13. For the relationship between product quality and the labor-intensity of the production process, see Lecraw, "Choice of Technology."

14. For more complete analyses of the development of technology in developing countries, see the following: Diana Crane, "Technological Innovation in Developing Countries: A Review of the Literature," *Research Policy*, 6 (Sept. 1977); Peter O'Brien and Jan Monkiewicz, "Technology Exports from Developing Countries: The Cases of Argentina and Portugal." Mimeographed, Geneva: UNIDO/IS 218, March 1981; Wells and Ghemawat, "Generation of Industrial Technology"; Lall, "Developing Countries as Exporters"; and O'Brien, "Third World Industrial Enterprises."

international transaction—for example, exports or licensing—or, going one step further, why they would become involved internationally at all. To take the last part of the question first, several motivations for international activities have been identified.

First, the market in the home country may have been quite small and quickly saturated. In order to continue to grow, these firms have either had to diversify into other products or begin to export some of their production to developed or developing countries. The incidence of firms with highly diverse products and activities in some developing countries suggests that some firms have followed a growth strategy of conglomerate diversification within their own local market.[15] Other firms, particularly those that had successfully modified their product and process technology so that they were efficient producers given factor and product market conditions in their local markets, began to export their products, often to other developing countries at lower levels of economic and industrial development—that is, they have exported to downstream countries that had a demand for the type of products they could produce efficiently and at low cost.

As these downstream countries have developed over time, however, local firms have often developed the

interest and capability of producing similar products, and their governments have fostered this development through investment incentives and protection from imports. In response to this threat to their export market, a threat made especially acute due to the price-sensitive nature of the demand for their products, some exporters from developing countries have been motivated to undertake FDI to protect their markets. Their proprietary product and process technology has given them a competitive advantage over both MNEs and locally owned firms in other low-wage, low-income markets. Firms that were motivated to undertake FDI for these reasons have typically started as producers of import substitutes for their home-country markets and have invested in import-substituting industries in host country markets. Through their FDIs, they have transferred product and process technology that was appropriate to the factor conditions, product demands, and market conditions in the host country.[16]

In many developing countries, however, government tariff and nontariff barriers to trade, competition policies, and the industrial structure have allowed firms to earn high profits in their home country over the long run. Firms in these countries generally have not had the incentive to adjust their product and process technology to meet local factor and demand conditions. These firms have not developed the competitive advantage in costs or products that was necessary to export

15. In India, government restrictions on market share, both in the base industry of large firms and in their investments in local industries outside their base industries, were the major incentive for FDI by the firms in Cordeiro's sample; see Carlos Cordeiro, "The Internationalization of Firms: A Case for Direct Foreign Investment from a 'Less Developed Country'" (undergraduate honors thesis, Department of Economics, Harvard College, 1978), Table XIV, p. 53.

16. The reason why they transferred their technology internally through FDI rather than via the market for product and process technology will be described later in this article.

or to undertake FDI, nor, given high profits at home, have they had the incentives to do so.[17]

A second group of MDCs have been motivated to undertake FDI, again usually in downstream countries, in order to remain competitive in their export markets. These firms initially exported labor-intensive, low-cost products, usually to high-income, high-wage countries. Their initial competitive advantage lay in their access to low-cost labor in their home countries and, less frequently, to their access to locally available raw materials. They also have developed proprietary skills and technology to produce labor-intensive products, using flexible, small-scale production technology and sometimes locally available inputs. Typically, however, the product and process modification and innovation of these firms has been less than for the first group of firms described previously, since their products were usually sold in developed countries, they produced fewer products at greater volume, and they often have had easier access to imported machinery and inputs due to the outward-looking nature of the development policies of their home country.[18]

As the home countries of these firms have developed, wage and other factor costs have risen, and their products have lost competitive advantage in export markets. In some cases, their export markets in developed countries have also been threatened by tariff and nontariff barriers to trade aimed specifically at products originating from their home countries. In response to these threats to their export markets, these firms have made downstream investments both to utilize low-cost labor and other cheap factor inputs in these countries and to circumvent trade barriers applied to their products.[19]

These MDCs have typically formed joint ventures with local entrepreneurs in the host country.[20] Their share of the equity in the joint venture has often been represented by new or used equipment shipped from the home country. Wells has noted that some textile firms in Hong Kong have taken equipment that was no longer efficient in relation to Hong Kong's high labor costs,

17. See Lecraw, "Choice of Technology," for the factors that have caused firms in Thailand to choose appropriate technology.

18. For an excellent study of the competitive advantage and the choice of technology of export-oriented MDCs that made FDIs in the Philippines and Mauritius, see Vinod Busjeet, "Foreign Investors from Less Developed Countries: A Strategic Profile," (DBA thesis, Harvard Business School, 1980).

19. The strong link between trade and FDI by MDCs has been noted by Louis Wells, Jr., "Multinationals from Latin American and Asian Developing Countries: How they Differ." Mimeographed, Harvard Business School, November 1980. Wells goes one step further and concludes that relatively greater numbers of MDCs based in Asian countries compared with those based in Latin America may have been due to the more export-oriented growth of many Asian countries.

20. This is not the place to undertake an analysis of the determinants of the type of equity position taken by MNEs or MDCs in their subsidiaries abroad (100 percent ownership, majority-ownership, or a minority position). MNEs based in the United States and Europe typically have taken 100 percent or majority ownership; MNEs based in Japan typically formed joint ventures; and MDCs have typically taken minority positions. The equity structure of subsidiaries in developing countries would seem to be the result of a complex process based on the bargaining power and desires of the foreign investor, the bargaining power and the desires of local entrepreneurs and the host government, and —arguably—government policies in the home country and the empathy the foreign investor feels for the goals and aspirations of host countries.

repainted it, labeled it as new, and shipped it to their subsidiaries abroad as part of their equity investment.[21] These firms also supply the joint venture with their expertise in using labor-intensive equipment and managing labor-intensive production facilities and with their access to export markets in developed countries.

Despite the differences in the characteristics of the firms in these two groups of firms and the differences in their competitive advantage and motivation for undertaking FDIs, they have often chosen to transfer their technology internally within the firm through FDI rather than by outright export of machinery or by licensing their product and production technology. In general, the market for technology is highly imperfect and involves the transfer of a complicated bundle of production and management skills as well as equipment and technical services, both at the time of the initial transfer and on an ongoing basis. Knowledge of the costs, benefits, risks, and opportunities of a technology package, both at the time of transfer and over time, is often highly asymmetrical between buyer and seller. This characteristic of technology transfer often makes pricing technology in the market extremely difficult and unsatisfactory to both the potential buyer and the seller. This problem is often compounded for firms based in developing countries, since their proprietary production technology may be embodied in the managers and workers who are experienced in working with that technology and

are only available within the firm in the home country. These skills often cannot be transferred to the host country by manuals, user guides, or short training courses for managers or production supervisors. The proprietary technology of firms in developing countries is often simpler than that of firms in developed countries in terms of equipment, but more complex in terms of the skills and procedures required of the managers and workers who use the technology. If these complex managerial and production skills are not transferred along with the equipment, much of the value of the technology is lost. This characteristic of the proprietary technology of MDCs has often made FDI the most efficient means by which the technology package could be transferred to the host country.

The low cost of sending managers and workers abroad for MDCs compared to MNEs has also increased the use of FDI as a means of transferring technology. Typically subsidiaries of MDCs, at least initially, have employed a higher proportion of workers and managers from the home country and have engaged in more extensive training programs in order to transfer their "people skills" to workers in the host country than have the subsidiaries of MNEs. This transfer of technology embodied in managers and production workers has sometimes worked the other way. Some MDCs have used their subsidiaries abroad to develop more general management skills in production managers from their firm in the home country through exposure to the wide range of problems they had to face in managing subsidiaries abroad. After they gained this experience, they were brought home in positions of increased responsibility.

21. Louis Wells, Jr., "Foreign Investors from the Third World: The Experience of Chinese Firms from Hong Kong," *Columbia Journal of World Business* (Spring 1978).

Other motivations for FDI by MDCs include diversification of political and economic risk across markets in several countries; export incentives for capital equipment by home-country governments in the form of direct subsidies and access to import licenses and foreign exchange;[22] desire to secure raw material sources; development of markets and marketing and distribution channels in the host country; investment incentives in the host country; acquisition of technology through purchase of firms abroad; and utilization of excess production, technical, and management capacity at home.[23] Whatever the motivation for FDI, however, MDCs have relied on the use and transfer of proprietary product and process technology in the host country developed in their home country that has given them the competitive advantage necessary to undertake FDI.[24]

22. MDCs that have been motivated to undertake FDI by incentives to export capital equipment have typically used the value of the capital equipment as their share of the equity in a joint venture abroad.

23. For data on these points, see the following: Cordeiro, "Internationalization of Firms," Table XIV, p. 53; Lecraw, "Foreign Direct Investment," Tables III, V, VI; idem., "Internationalization of Firms from LDCs," Table 4; and Louis Wells, Jr., and V'Ella Warren, "Developing Country Investors in Indonesia," *Bulletin of Indonesia Economic Studies* (Mar. 1979).

24. An analysis of the complex factors that have led to FDI by MDCs is developed by John Dunning, "Explaining Outward Direct Investment of Developing Countries: In Support of the Eclectic Model," in *Multinationals from Developing Countries*, eds. Krishna Kumar and Maxwell McLeon (Lexington, MA: D.C. Heath, 1981). For descriptions of firms in Latin America, see Eduaro White, "The International Projection of Firms from Latin American Countries," in ibid. For Hong Kong, see the following: Louis Wells, Jr., "Foreign Investors from the Third World: The Experience of Chinese Firms from Hong Kong," *Columbia Journal of World Business*

In general, MDCs did not undertake FDI in more developed, more industrialized countries, since market conditions in their home country did not motivate them to develop technology that gave them the competitive advantage necessary to compete via FDI in upstream countries.[25]

IMPACT OF MDCs ON HOST AND HOME COUNTRIES

The impact of FDI by MDCs on both the host and home country may have been quite different from that of other MNEs. This difference was largely due to the type of technology they transferred as part of their investment. Their technology was often more appropriate for small-scale production, was more labor-intensive, utilized a higher

(Spring 1978); and Edward K.Y. Chen, "Hong Kong Multinationals in Asia: Trends, Patterns and Objectives," in *Multinationals from Developing Countries*, eds. Kumar and McLeon (Lexington, MA: D. C. Heath, 1981). For Korea, see Sung-Hwan Jo, "Overseas Direct Investment by South Korean Firms: Direction and Pattern,"in ibid. For Southeast Asia, see Lecraw, "Foreign Direct Investment," idem., "Internationalization of Firms from LDCs," and idem., "Intra-Asian Direct Investment: Theory and Evidence from the ASEAN Region," *UMBC Economic Review*, 16(2) (1980). For Indonesia, see Wells and Warren, "Foreign Investors from the Third World." For Taiwan, see Wen-Lee Ting and Chi Shive, "Direct Investment and Technology Transfer from Taiwan," in *Multinationals from Developing Countries*, eds. Kumar and McLeon (Lexington, MA: D.C. Heath, 1981). For India, see Usha Dar and Pratap Dar, *Investment Opportunities in ASEAN Countries* (New Delhi: Sterling Publishers Pvt. Ltd., 1979) and Cordeiro, "Internationalization of Firms."

25. There have been exceptions to this generalization: a few firms based in Taiwan and India have made FDIs in manufacturing facilities in the United States and Europe. When Pernas—of Malaysia—acquired Sime Darby and Guthrie, it obtained subsidiaries in the United Kingdom and Australia.

proportion of locally produced inputs of raw materials and ancillary equipment, was more flexible between processes and products, operated at higher capacity utilization, and—at least initially—required a greater proportion of workers and managers from the home country than did the technology transferred by other MNEs. Although the data are very sketchy, the technology transferred by MDCs would seem to have been more appropriate to the factor costs and demand conditions in other developing countries than that transferred by other MNEs.

The linkages to the local economy in terms of demand for locally produced inputs of raw materials and equipment and possibly the training of local workers and managers may also have been greater. The greater tendency of MDCs to make FDI using joint ventures may have enhanced this linkage effect. On the other hand, a significant proportion of FDI of MDCs has been made to forestall potential investment by entrepreneurs in the host country who might have been able to produce for the local market by themselves. Moreover, much of the FDI by MDCs in Asia and Africa has been undertaken by entrepreneurs belonging to ethnic groups that already played a significant role in the economies of host countries, and the FDI may have served to increase or preserve this position, sometimes contrary to implicit or explicit government policy. The strategic profile of MDCs that produce low-quality, unbranded, technologically old, low-cost, price-sensitive products in host country markets has tended to reduce their impact on the competitive practices and supranormal profits earned by MNEs in developing countries.[26] On balance, however, FDI and technology transfer by MDCs would seem to have had a positive impact on the economies of the host countries and to have served as a useful complement, if not supplement, to FDI by other MNEs.

Technology transfer by MDCs has also had an impact on the economies of the home countries of the MDCs. Although the effects are more difficult to trace and substantiate using currently available data, it would seem that FDI by MDCs has increased the profitability of undertaking product and process modification and innovation by firms in developing countries, increased exports of equipment that embodied technology produced at home, increased the amount of research and development undertaken in the home countries, and expanded the opportunities and demand for skilled workers and professionals with engineering and science degrees. In sum, transfer of technology by FDI by MDCs would seem to have had a positive—although as yet quite small—influence on the opportunities available to firms in developing countries and their capabilities to produce increasingly sophisticated products using increasingly sophisticated processes.

The future of technology transfer by MDCs via FDI is not clear. Based on fragmentary data, there seems to

26. For an analysis of business strategy and strategic groups applied to MDCs, see Busjeet. For a statistical analysis of the impact of FDI by MDCs on industry structure, competitive practices, and profits in the ASEAN region, see Lecraw, *Structural and Competitive Practices of Transnational Corporations in the ASEAN Region* (New York: U.N. Center on Transnational Corporations, 1979).

have been a trend toward increased FDI by firms based in developing countries in absolute amount and possibly relative to FDI by MNEs. In order for this trend to continue in the future, however, firms based in the higher-income developing countries must continue to upgrade and develop their product and process technology at home and subsequently transfer it to their subsidiaries abroad. Wells[27] has noted that when some firms based in Argentina failed to follow this path, their subsidiaries abroad gradually became more independent of their parents and in some cases ended their relationship entirely.[28] It would seem, however, that FDI by MDCs based in the more industrialized of the developing countries will have a continuing—and possibly increasing—role to play in filtering, modifying, and developing technology that is appropriate to the factor market and demand conditions in downstream developing countries and that they will transfer this technology through direct investment.

27. Wells, "Multinationals from Latin America and Asian Developing Countries."

28. This evolution may be atypical, however, and may have been due to the general stagnation and decline of the manufacturing sector in Argentina as a whole; see O'Brien and Monkiewicz, "Technology Exports from Developing Countries."

From Dependency to Interdependency: Japan's Experience with Technology Trade with the West and the Soviet Union

By STEPHEN STERNHEIMER

ABSTRACT: As the debate over the merits and demerits of technology trade between East and West gathered force in the 1970s, an awareness slowly dawned that the issues at stake touch the fortunes not only of the United States but those of her allies as well, including Japan. Japan's own success with technology imports, beginning only a few decades ago, gives grounds for new fears regarding future competitiveness on the part of a reinvigorated Soviet economy. Further, Japan's raw-materials vulnerability and the importance of export markets to her economy have led some to suggest that Japanese technology trade with Communist nations might leave her leadership politically vulnerable to pressure from Moscow and elsewhere. As this article demonstrates, such fears are largely groundless. The Japanese and Soviet approaches to technology transfer for economic development display far more differences than similarities. Likewise, in terms of a potential Japanese export dependency on Soviet markets, Tokyo remains more critical to Moscow than the reverse. From the perspective of Japanese import dependency, especially in the energy sector, a review of the statistical picture that can be drawn up for Japanese-Soviet energy cooperation in Siberia reveals that the Soviet contribution to Japan's energy balance, even in 1990, will be quite minimal. The real issue, this article concludes, arises from the need for the United States to develop a policy for the control of technology trade between East and West that will reflect the needs and perspectives of all Western allies, rather than those of the United States alone.

Stephen Sternheimer, political scientist at Boston University and a research associate of the Russian Research Center, Harvard University, was trained as a Soviet specialist—B.A. Harvard University, Ph.D. University of Chicago. He recently served as a consultant to the U.S. Department of Defense on technology transfer to China and to the Office of Technology Assessment, U.S. Congress, on Japanese-Soviet technology trade. His published writings concern Soviet urban politics, Soviet public administration, and technology trade between Japan and the USSR. Currently he is involved in a study of retirement, aging, and work in Communist systems funded by the National Institute on Aging, U.S. Department of Health and Human Services.

E VEN casual review of the bur- geoning literature on East- West technology transfer soon leaves an intelligent reader convinced that he has entered a looking- glass worthy of Lewis Carroll's best efforts. A recently published *tour d'horizon* by the Organization for Economic Cooperation and Develop- ment (OECD) catalogues a plethora of technology transfer (TT) issues, ranging from historical trends and definitional-measurement prob- lems through questions of govern- ment policies, economic and strategic implications, the innova- tion capabilities of recipient economies—especially centrally planned economies—and the var- ious costs and benefits accruing to the sending nation.[1] Some analysts stress that the industrialized West- ern (IW) nations stand to gain sub- stantially from such transfers, as embodied in sales of machinery and equipment, turnkey plants, and licensing arrangements, by virtue of longer production runs, higher lev- els of output, and additional funds for domestic research and develop- ment (R&D) efforts.[2] They argue for the importance of trade with the Eastern European bloc and the Soviet Union as America's "greatest opportunity" for expanding exports;

NOTE: Much of the research on which this article is based is the result of two studies carried out under contract for the Office of Technology Assessment, United States Con- gress, in 1979 and 1980-81. The views expressed here are those of the author and in no way represent those of the contracting agency or the U.S. government. Footnoted published material was supplemented with interviews with Japanese and American pol- icy officials and businessmen in Tokyo, Washington, New York, and Cambridge.

1. Eugene Zaleski and Helgard Weinert, *Technology Transfer Between East and West* (Paris: OECD, 1980), pp. 12-14.
2. Ibid., pp. 274-81.

trade represents a desirable thrust for American foreign economic pol- icy that should not fall prey to a "don't-sell-them-anything" attitude on the part of those military special- ists preoccupied with strategic mat- ters.[3]

Another school of thought, pro- ceeding from the assumption that "'world leadership and technology leadership are inseparable [such that] a third-rate technological nation is a third-rate power, politi- cally, economically, and socially'," claims that industry in the United States must "be more careful about know-how sales to state-owned com- petitors, which create competition in Communist economies." In part this is attributed to the fact that it is more expensive to be a leader than a follower in the realm of technology development.[4] In addition, it is adduced, Western nations must con- tinue to rely on technological super- iority to offset the Soviet Union's advantage in quantitative military power. In the eyes of such individu- als, what the West is doing through TT is neither more nor less than sel- ling the USSR the rope with which the West will be hung, as Lenin pro- phesied over half a century ago.[5]

3. See *Multinational Corporations and United States Foreign Policy*. Hearings before the Subcommittee on Multinational Corporations of the Committee on Foreign Relations, U.S. Senate (Washington: GPO, 1977), p. 143 (testimony of Mr. Donald Green); pp. 75-76 (testimony of Mr. William Casey); Marshall Goldman, "US Policies on Technology Sales to the USSR," in NATO, Directorate of Economic Affairs, *East-West Technological Cooperation* (Brussels: NATO, 1976), pp. 103-4.
4. *Multinational Corporations*, pp. 244- 45 (testimony of J. Fred Bucy); J. Fred Bucy, "Technology Transfer and East-West Trade: A Reappraisal," *International Security*, 5(3):137 (Winter 1980-81).
5. Carl Gershman, "Selling Them the Rope: Business and the Soviets," *Commen- tary*, 67(4):35-45 (Apr. 1979).

Thus we see that the issues with which the informed observer must grapple in any consideration of the implications of East-West TT are numerous and complex. In particular, the question of future competitiveness on the part of Communist nations—issue number one—needs to be evaluated in light of the Japanese experience with TT from Western sources during the two decades that followed the close of World War II, particularly with respect to the substantial export advantages Japan today enjoys by comparison with the rest of the industrialized world. To this we would add issue number two—the possibility of an emerging "import dependence" between Western technology sellers and Communist purchasers, especially given the countertrade—or barter—nature of the majority of such transactions.

There is the related problem—issue number three—of a possible export dependency, either for the United States or, more likely, for such U.S. allies as Japan. According to this scenario, TT increases the opportunities for Soviet political leverage over Western policymakers by virtue of the economic "bait" that it provides. Some historical precedent exists in the form of the kind of "friendship trade" that China foisted on Japan during the late 1960s and early 1970s.[6] As one analyst of the "hoisted-by-their-own-petard" school of thought—now on the staff of the U.S. mission to the UN—contends, "capitalism is more efficient than communism, but if the very efficiency is used to sustain and fortify the enemies of free society, does this not, in the words of Sey-

mour Martin Lipset, constitute the 'ultimate failure of capitalism.'?"[7]

These, in capsule form, constitute the TT issues that this article addresses from the vantage point of Tokyo. Turning first to Japan's own experience with technology imports, we will ascertain whether and to what extent Japan's current export competitiveness, as it derives from earlier patterns of TT, offers us a taste of what the future might bring vis-à-vis trading relations with the Communist world. Second, we will briefly survey Japan's technology trade with the Soviet Union as a case study of the extent and means whereby U.S. allies contribute to the process of East-West—or, more accurately, West-East—TT. Finally, we will examine current Japanese technological contributions to Soviet energy development in terms of such issues as fuel-import dependency, export-share dependency, and the ensuing opportunities that might be emerging for the Soviet Union to fragment the Western alliance politically. In each case, we will also concern ourselves with those kinds of factors that might shape Allied reactions to future U.S. efforts to regulate TT to Communist nations more stringently.[8]

JAPAN AS TECHNOLOGY IMPORTER AND LESSONS FOR EAST-WEST TRANSFERS

One of the issues that perennially surfaces in any discussion of East-West TT is a fear that, by virtue of current technology imports, Communist states will eventually be able

6. C.J. Lee, *Japan Faces China: Political and Economic Relations in the Post-War Era* (Baltimore: Johns Hopkins University Press, 1976), pp. 141-69.

7. Gershman, p. 45.
8. For a more detailed treatment of the dynamics of Allied responses, see Stephen Sternheimer, *East-West Trade and Technology Transfer: Japan and the Communist Bloc,*

to export products competitive with Western industries. This is particularly true for those sectors sensitive to import penetration and already deeply troubled owing to declining world demand, high domestic labor and materials costs, and a production infrastructure that is frequently outmoded in terms of the technological level of its capital equipment.[9] However, a preliminary analysis of trends evident by the mid-1970s reveals that centrally planned economies (CPEs) "provided a very stable 3-3.1% of all IW imports in sensitive import sectors" —textile fibers, chemical elements, plastics, steels, manufactured fertilizer, clothing, footwear, and fabric products—despite an overall increase in CPE exports to the IW nations of some 70 percent.[10] Such findings suggest that in order to assess the future competitiveness of these economies, it would be useful to examine the Japanese experience with technology imports and the reasons why the Japanese were able to capitalize on these to gain a lead-

ing edge in certain export areas over the last decade.

In any review of Japan's success with her technology importation strategy, several features are worth noting. First, as Peck and Tamura have pointed out in a seminal study of the Japanese experience for the Brookings Institution, the initial period of technology purchases, 1952-60, was underlaid by a unified industrial policy administered by Japan's powerful Ministry of International Trade and Industry (MITI), sometimes facetiously labeled by the Japanese "the Ministry of Everything."[11] It was designed to make Japan self-sufficient in what economists term "intermediate material inputs," including chemicals, iron, and steel. Second, once some success in this area was achieved, MITI shifted quickly to favoring technology imports oriented to potential exports of consumer goods, accompanied by a liberalization of import controls, domestic R&D emphasizing improvements in previously imported technology, and initiatives to promote imports. This was designed to encourage competition, productivity, and innovation among Japanese purchasers, resulting in rapid movement to recoup production and capital investment costs. By

The Washington Papers, No. 76 (Beverly Hills: Sage Publications, 1980), pp. 69-79 (hereafter cited as Japan and the Communist Bloc); Angela Stent Yergin, East-West Trade and Technology Transfer: European Perspectives, The Washington Papers, No. 75 (Beverly Hills: Sage Publications, 1980) pp. 17-28, 51-53, 60-63, 84. The background to alliance strains is discussed in Takuya Kubo, "Security in Northeast Asia," in Asian Security in the 1980s: Problems and Policies for a Time of Transition, ed. Richard H. Solomon (Santa Monica: Rand Corporation, 1979) and Horst Menderhausen, Outlook on Western Solidarity: Relations in the Atlantic Alliance System (Santa Monica: Rand Corporation, 1976), esp. pp. 8-17 and 118-34.

9. Karen Taylor and Deborah Lamb, "Communist Exports to the West in Import Sensitive Sectors," in Issues in East-West Commercial Relations, Joint Economic Committee (JEC), U.S. Congress (Washington, DC: Government Printing Office, 1979), pp. 125-56 (hereafter cited as Issues).

10. Ibid., p. 133.

11. The discussion that follows draws chiefly on M. Peck and S. Tamura, "Technology," in Asia's New Giant: How the Japanese Economy Works, eds. Hugh Patrick and Henry Rosovsky (Washington: Brookings Institution, 1976), pp. 525-86; Lawrence B. Krause and Sueo Sekiguchi, "Japan and the World Economy," in ibid., pp. 398-402, 407-10, 421-22; Subcommittee on Trade of the Committee on Ways and Means, U.S. House of Representatives, High Technology and Japanese Industrial Policy (Washington, DC: Government Printing Office, 1980), pp. 10-18; Terutomo Ozawa, Japan's Technological Challenge to the West, 1950-74 (Cambridge: MIT Press, 1974), pp. 16-51, 67-80, 85-87.

the end of the 1960s, improvements in previously imported technology received high priority. Meanwhile, imports accounted for the vast majority of additional technology purchases, and firms invested substantial sums in indigenous research to improve on technologies already in their possession.[12]

Such an experience stands in sharp contrast to that of the USSR. Traditionally, the Soviet Union has done well competitively only in processes not noted for technological complexity in which central coordination of large amounts of capital and resources confers some kind of comparative advantage—for example, the fishing industry. While the Soviet Union has been able to enter foreign markets in sectors where technology is not particularly fast-moving, the immunity the Soviet economic system provides against the measurement of success in terms of product-by-product or firm-by-firm profitability means that the Russians have had only limited success in the sale of items such as watches, cameras, tractors, and automobiles. The system's inability to turn invention into production— and the structural divorce of Soviet research scientists from industry— constitutes yet another obstacle and a point of dissimilarity with Japan, where most R&D is privately financed and supervised. As a lead-

ing specialist on Soviet trade, Marshall Goldman, has noted, "Even in those areas where the Soviet Union seems to have an advanced technology and a reputation for the production of certain items, the Russians sometimes find it necessary to import technology, machinery, and advice, as for example in the coal, timber, and steel industries."[13]

Other reasonably "unique" characteristics of Japan's handling of technology imports abound. The Japanese have proven unusually adept at utilizing their quite sophisticated and complex information-gathering capacities, which are characterized by strong horizontal integrative ties in contrast to the vertical fragmentation of the Soviet system.[14] The structure of the "all-round trading companies" *(sogo shosha)*, with their intricate networks of manufacturing, financing, marketing, and trading linkages, constitutes another feature of the Japanese economy that has no parallel in contemporary Communist systems.[15] Likewise, investments in adaptive research have been considerable. By 1972, a MITI survey revealed that at least one-third of all R&D expenditures were targeted at modifying or improving technologies already imported. Again, it is difficult to find parallels in a system such as that of the USSR.[16]

More important, despite the fact that Japan's receipts for selling technology continued to lag behind her payments for technology already imported, a review of the ratio of re-

12. Goldman, pp. 105-9; Thomas A. Wolf, "The Distribution of Economic Costs and Benefits in U.S.-Soviet Trade," in Joint Economic Committee, U.S. Congress, *Soviet Economy in a Time of Change*, vol. 2 (Washington, DC: Government Printing Office, 1979), pp. 329-32. The limits to Soviet ability to absorb technological efficiency via TT are skillfully analyzed in Phillip Hanson, "The Diffusion of Imported Technology in the USSR," in NATO, *East-West Technological Cooperation*, pp. 144-49.

13. Goldman, p. 108.
14. Ezra Vogel, *Japan as Number One: Lessons for America* (New York: Harper Colophon, 1979), pp. 27-52.
15. Ibid., pp. 131-57; A. K. Young, *The Sogo Shosha: Japan's Multi-National Trading Companies* (Boulder: Westview Press, 1979); Krause and Sekiguchi, pp. 389-97.
16. Peck and Tamura, p. 542.

ceipts to new payments reveals that by 1973, Japan had become a net exporter of technology to other nations.[17] This stands in sharp contrast to the Soviet case where, despite a 2 to 1 advantage in the number of licenses sold to the United States as opposed to the number purchased, the ratio of Soviet payments to receipts—for the United States alone—stood at almost 10 to 1 by the mid-1970s, thereby making the Soviet Union a net importer by a factor of at least 10.[18]

Once R&D for military purposes is excluded, by 1971, Japan spent only slightly less of its GNP—1.6 percent—on nonmilitary R&D than did the United States or West Germany—1.9 percent—or Britain—1.7 percent.[19]

Thus, whereas the purposive nature of policymakers' approach to TT, plus the concern for enhancing economic growth through TT, displays numerous parallels between nations as diverse as Japan and the Soviet Union, the sheer number, complexity, and imaginativeness of Japanese strategies for importing and then exploiting technology make its experience largely unique.

JAPAN AS A SELLER IN
TECHNOLOGY TRANSFER TO
THE SOVIET UNION

The role played by Japan as a source of technology imports for

Communist nations has begun to command increasing attention after a prolonged period of benign neglect.[20] In part this stems from the growing debate between two points of a view vis-à-vis the merits and demerits of technology trade. One regards commerce among nations as a necessary, if not sufficient, condition for achieving peaceful relations with the Communist world; the other emphasizes that too-expensive commercial and technological intercourse with a potential Communist adversary can only strengthen the economic and military capabilities of the trading partner in ways that endanger Western security.[21] Interest in the Japanese position in terms of the debate has risen out of the growing tensions evident within the Western alliance, circa 1975-80, over the issue of the proper role for commerce and economic interdependence—as opposed to military confrontation and economic warfare—in relations with the East.[22] Such strains have come to the fore in the wake of such Washington initiatives as efforts to restrict the flow of "critical technologies" to Communist nations, 1976-present; the Afghanistan-inspired embargo

17. Ibid., p. 541. See also Sternheimer, *Japan and the Communist Bloc*, pp. 50-52 and *Outline of the White Paper on Science and Technology* (Tokyo: Foreign Press Center, 1977), pp. 11-12.

18. John Kiser, "Report on the Potential of Technological Exports from the Soviet Union to the United States," unpublished report prepared for the National Science Foundation (NSF) and Department of State, 1977, pp. iii, 74.

19. Peck and Tamura, p. 533 (Table 8.4).

20. All the same, with the exception of sections in the recent report of the Office of Technology Assessment, U.S. Congress, *Technology and East-West Trade* (Washington, DC: Government Printing Office, 1980), the long-awaited OECD report, *Technology Transfer Between East and West*, by Eugene Zaleski and Helgard Weinert (Paris: OECD, 1980), and Stephen Sternheimer's monograph, *Japan and the Communist Bloc*, no other systematic treatments of the subject have appeared in English.

21. Office of Technology Assessment (OTA), U.S. Congress, *Technology and East-West Trade* (Washington, DC: Government Printing Office, 1980), p. 3. For a more detailed summary of the various viewpoints, see Zaleski and Weinert, Chapter 4 (part II).

22. OTA, pp. 12-13; Bucy, pp. 147-50.

on "high tech" exports to the USSR in 1980; and the current debate in the United States over the wisdom of West European and Japanese participation in the Yamburg pipeline project that would utilize Western technology and capital to bring Soviet natural gas to Western Europe.[23] Against such a backdrop, it is useful to investigate the Japanese position on such matters as the control of TT to the USSR, as well as Japan's contribution to East-West technology trade as compared with that of other IW nations.

The legal framework for control over TT to the Communist bloc in Japan is provided by the Foreign Exchange and Foreign Trade Control Law of 1949, as supplemented by periodic revisions of the lists of commodities subject to such regulation. This law embodies a presumption of license rather than restraint—in contrast to various export administration acts in the United States prior to 1979—and it aims at the normal development of foreign trade between Japan and all nations, subject to an unspecified degree of government intervention.[24] Amendments to the law promulgated in 1980 underscored the importance of the notion of "freedom to trade in principle," chiefly by liberalizing government authority over

capital transactions and revising trade settlement procedures. This moved Japan even further away from the kind of politically inspired orchestration of technology trade with the USSR that Western analysts such as Samuel P. Huntington —formerly a member of the National Security Council (NSC) and currently director of the Center for International Affairs at Harvard—have recommended.[25]

Export controls are not regarded in Tokyo as appropriate instruments for foreign policy, economic warfare, or safeguards of national security to the same extent as is true in the United States. The list of commodities subject to the provisions of the law was considerably shorter than Washington's list as late as 1978.[26] Conversely, the five-year trade agreements that Japan has concluded with the USSR periodically since 1966, with the latest one negotiated in the late spring of 1981, are used to help grease the wheels of Tokyo's licensing bureaucracy. They also serve to protect Japan against Soviet dumping and to ensure the absence of the kind of conflict over TT between government and business—about what can and cannot be legitimately traded— that seems to prevail in the United States.

In terms of the amounts and kinds of technology that Japan has provided to the Soviet Union in the course of the past decade, the pic-

23. For a discussion of the Yamburg pipeline project and its technological specifications as well as fuel-dependency ramifications, see OTA, "Soviet Energy and Western Technology" (Washington, DC: Government Printing Office, 1981); Stephen Sternheimer, "Japanese-Soviet Energy Cooperation: Dependency, Interdependency or Vulnerability?" unpublished final report submitted to OTA, April 1981, pp. 11-14 (hereafter cited as "Japanese-Soviet Energy Cooperation," Final Report); New York Times, 19 November 1980, p. D3; Soviet Business and Trade, 15 February 1981, p. 6.

24. R. Baker and R. Bohlig, The Inter-

national Lawyer, No. 1:163-81 (1967); Nisso Too Boekikai, Nisso Boeki Yoron (Tokyo: Nisso Too Boekikai Jimukiuku, 1959), pp. 292-93.

25. Digest of Japanese Industry and Trade, No. 144:37 (Apr. 1980); S. P. Huntington, "Economic Diplomacy," Foreign Policy, 32:65-73 (Fall 1978).

26. Sternheimer, Japan and the Communist Bloc, pp. 13-16; idem, "Japanese-Soviet Energy," pp. 98-106.

ture that emerges is a complex one. Overall, the USSR is not now and has never been a major Japanese trading partner; the 2 percent of Japanese imports she absorbed prior to World War II and the 2.5 percent of all Japanese imports the USSR provided has remained almost constant right up to the present time, 1981.[27] The sheer dollar value of such trade, however, has grown enormously, by a factor of 270 with respect to Japanese exports to the USSR, 1957-78, and by a factor of 117 with respect to Japanese imports—$2.5 billion and $1.4 billion, respectively, 1978.[28] However, a list of Japan's "best 20" trading partners for 1977 shows that the USSR ranked only eleventh as an export market and thirteenth as a source of imports, behind not only other Western nations but also the People's Republic of China—tenth and eleventh places, respectively.

In terms of actual transfers, Japan has exported turnkey plants to the Soviet Union in sectors such as chemicals production, fertilizer manufacture, and metallurgy even as it has provided machinery for mining, construction, and a wide variety of other sectors. Such items, however, do not necessarily contain high technology or state-of-the-art technology, and many of these are

excluded from the list of high tech items used by the U.S. Department of Commerce for discussions of export controls.[29] In terms of TT through licensing, one source suggests 200 such agreements in operation between Japan and the Soviet Union by the mid-1970s—which amounts to but 8.5 percent of all such arrangements between IW nations and the countries of Eastern Europe and the Soviet Union.[30]

If other indicators of TT to the USSR, such as "machinery and equipment" sales, are used, Japan's contribution to the Soviet Union's stockpile of Western technology appears large in terms of all Japanese exports to the USSR—which are themselves small in terms of the value of trade involved—but insignificant by comparison with Japanese exports to other world regions.

Using other definitions of "technology"—such as "technology-intensive products," a catalogue developed by Raymond Mathieson, and "high technology" items, a catalogue developed by the U.S. Department of Commerce—we find that Japan plays a major role as a source of TT for the USSR in selected areas. Mathieson's criteria, applied to Japanese-Soviet trade statistics for 1973, reveal that Japan accounted for the major share of Soviet imports globally in three sectors—electrolytic industry anodes, chemical equipment, and electronics equipment—

27. Roger Swearingen, *The Soviet Union and Post-War Japan: Escalating Challenge and Response* (Stanford: Hoover Institution, 1978), pp. 143-55; Kazuo Ogawa, "Economic and Trade Relations Between Japan and the Soviet Union, China, and the Socialist Countries of Europe," unpublished paper for Soren Too Boekikai, Tokyo, May 1979.

28. Sternheimer, *Japan and the Communist Bloc*, pp. 27 (Table 1), 32-33 (Figures 1 and 2). Overall, Japanese exports to the Communist world accounted for 6 percent of her exports globally in 1977, as compared to 24 percent to the United States, 11 percent to the European Economic Community (EEC), and 21 percent to developing nations in Asia. MITI, *White Paper on International Trade, 1978: Summary* (Tokyo: MITI, 1979), appendix, pp. 76-77.

29. Hedija Kravalis et al., "Quantification of Western Exports of High Technology Products to Communist Countries" in JEC, *Issues*, pp. 44-45. For an overview of the definitional diversity that has sprung up around efforts to define "technology trade," see T. M. Podolski, "Evolution of East-West Technology Transfer Channels," in NATO, *East-West Technological Cooperation*, pp. 119-32.

30. Raymond Mathieson, *Japan's Role in Soviet Economic Growth* (New York: Praeger, 1979), pp. 154-55, 158, 233-34; Phillip Hanson, "Western Technology in the Soviet Economy," *Problems of Communism*, 17:22-23 (Nov.-Dec. 1978).

and for major shares of Soviet imports from IW nations in these sectors as well as in metal-cutting instruments, machine tools, and hoisting gear.[31]

West Germany, however, occupied first place among all IW nations in 1977 as a source of Western technology and was more important than Japan globally as well. The United States was in direct competition with Japan for several categories of products. When a rank ordering of Soviet supply sources for high tech commodities is constructed for 1972-77, Japan occupies the position of leading supplier for two commodities, as opposed to eleven for Germany, five for France, two for Britain, and two for the United States.[32] However, if the more aggregate—and some claim, more accurate—measure of "whole plant exports" is utilized in examining TT between Japan and the USSR, then the USSR again surfaces as a major Japanese partner, accounting for one-third of all Japanese sales in 1976, versus one-fifth to Latin America and one-fourth to East Asia.[33]

Such statistical pictures, however tedious the details may appear, are virtually necessary if we desire to develop accurate and dispassionate assessments on the basis of which to make policy judgments for the forthcoming decade. Clearly the United States no longer acts as "sole source" supplier for the USSR in many areas, just as Japan, in others,

occupies second place behind West Germany. This suggests that any successful effort on the part of the United States to regulate TT to Communist nations must take careful account of just whose ox will be gored by specific kinds of controls.

Second, the statistics suggest that on the balance, Japan as a source of Western technology imports is more important to the USSR than the Soviet Union is as a market for Japanese exports, technology-intensive or otherwise. From this perspective, the probability of the Kremlin being able to turn the trade linkage to political advantage owing to some ill-defined "dependency" by Japanese industry on Soviet purchases appears farfetched. Of course, there is always the chance that decision makers in Moscow could prove completely impervious to the fact that a drive for political influence could kill the goose that lays golden eggs.

JAPANESE TECHNOLOGY
TRANSFERS AND SOVIET ENERGY:
NEW ISSUES FOR THE 1980s

In order to focus more closely on some of the issues associated with East-West technology trade, it is useful to examine in some detail one aspect of Japanese-Soviet technological cooperation—for example, the development of Siberia's energy resources.[34] Many Western scholars have underscored the importance of rapid Siberian development for the overall Soviet energy balance and the health of the Soviet economy in

---

31. Sternheimer, *Japan and the Communist Bloc*, pp. 38-39 (Table 8).

32. Ibid., p. 48.

33. Calculated on the basis of data from JETRO, *Japan's Plant Exports*, No. 11 (Tokyo: Japan External Trade Relations Organization, 1977), pp. 3-4. The advantage of plant exports as a channel for TT are described in A. Hausman, "East-West Technology Cooperation, Likely Transfer Patterns, (1976-1980)," in NATO, *East-West Technological Cooperation*, p. 336 and Podolski, p. 130.

34. For a more comprehensive treatment, see Stephen Sternheimer, "Japanese-Soviet Energy Cooperation: Dependency, Interdependency or Vulnerability?", Interim and Final Reports to OTA, January 1981 and April 1981 (hereafter cited as "Japanese-Soviet Energy Cooperation," Interim Report).

general.[35] Other sources stress the critical role Japan is likely to play in this respect, by virtue of its geographical proximity, a foreign policy that equates political security with economic interdependence, and a desperate need for increased stores of fuels and raw materials to feed its burgeoning industrial economy.[36]

In comparison with other IW nations as suppliers of energy-related technology (ET) to the Soviet Union, Japan's performance, 1975-79, has been erratic, accounting for roughly 20 percent of all such IW exports to the USSR in 1975, 38 percent in 1976, 26 percent in 1977, and 37 percent in 1978 and 1979.[37] Japan's technological contribution made a particularly strong showing among IW sources for the USSR in areas such as "tubes, pipes, and fittings," 31-53 percent for 1975-79; "air pumps, vacuum pumps, and air or gas compressors," 40 percent; "special purpose motor vehicles and cranes, etc.," 61 percent; "electrical measuring or automatically controlling instruments," 52 percent; and "floating structures, including drilling platforms," 54 percent. However, in other areas—coal mining equipment, surveying instrumenta-

tion, reciprocating pumps, and gas, liquid, or electricity production meters—the Japanese share of IW exports to the USSR has been minimal.

Unfortunately, such figures are by no means adequate as a measure of Japan's contribution, or lack thereof, to Soviet energy development. Only a fraction of the machinery exported in each case may have an energy-related function, and there is no reason to believe that the energy-related proportions are constant in the trade statistics of different nations. In any event, as a supplier of ET to the USSR among all IW nations, 1975-79, Japan ranks second after West Germany, followed by Italy, France, and the United States.[38]

What this suggests, among other things, is that, as in the case of technology trade in general, any U.S. efforts to curb energy technology exports to the USSR in retaliation for Soviet adventurism abroad—Afghanistan, the Middle East—or as a fee extracted for good behavior would make great demands on Western alliance solidarity and would extract larger costs from America's alliance partners than the price that American firms and trading interests would be called upon to pay. In such a setting, resistance in Tokyo, Paris, Bonn, Rome, or London will quickly dwarf political repercussions in Washington.

From the perspective of the relative market-share importance of the USSR as a customer for Japan's energy-related technologies, we find that Soviet economic opportunities for politically motivated leverage are rather small. Roughly speaking, Japan is about twice as important to the USSR as a supplier of energy-related technology among IW nations as the USSR is to Japan as a purchaser of ET on a global scale.

35. Marshall I. Goldman, *The Enigma of Soviet Petroleum: Half Empty of Half Full?* (London: Allen and Unwin, 1980): NATO, Directorate of Economic Affairs, *Exploitation of Siberia's Natural Resources* (Brussels: NATO, 1974); Leslie Dienes, "The Soviet Energy Policy," in *Soviet Economy in a Time of Change*, vol. 1, Joint Economic Committee (Washington, DC: Government Printing Office, 1979), pp. 196-229; Martin Spechler, "Regional Development in the USSR," in ibid., pp. 141-63.

36. Swearingen, pp. 86, 121-42; Young C. Kim, *Japanese-Soviet Relations*, The Washington Papers, No. 21 (Beverly Hills: Sage Publications, 1974), pp. 55-75; Saburo Okita, *Japan in the World Economy* (Tokyo: Japan Foundation, 1975), pp. 187-95.

37. Calculated on the basis of U.S. Department of Commerce tables provided to OTA, November 1980; Sternheimer, "Japanese-Soviet Energy Cooperation," Interim Report, pp. 2-4 (Tables 1 and 2).

38. Ibid., p. 8 (Table 3).

As in the previous case, the opportunities such a situation presents for translating economic ties into political pressures by Moscow seem remote.

Similarly, the import—fuel—dependency side of the picture gives few grounds for alarm. This holds despite the concern of some analysts that Western nations are about to trade their liberal democratic souls for a mess of Communist pottage in the form of oil, coal, or—currently—natural gas. For the Japanese case at least, such a problem exists only in the eyes of the beholder.

What does the future hold? In contrast to the many uncertainties surrounding Japanese technology exports of various types, some rather concrete, high-probability predictions vis-à-vis fuel purchases from the USSR are possible, using MITI's energy supply-demand projections for the Japanese economy for 1990. By employing MITI's matrix, we can, for example, predict the impact that the various Siberian projects Japan is currently assisting—South Yakutia coal, Yakutia gas, and Sakhalin offshore oil and gas—will have on Japan's "energy dependency" if they come onstream as scheduled.

Imported coal, which will account for 15.6 percent of Japanese energy sources in 1990, calculated in tons of oil equivalents, will be mined in the USSR only in a very small part—that is, 6.2 percent of Japan's total coal imports. Likewise, whereas "best case" projections suggest that imported oil will constitute only one-half of Japanese fuel sources by 1990, less than one-third of one percent of this amount will be pumped from Soviet wells; this assumes a minimally commercially exploitable deposit off the coast of Sakhalin, which no one in Tokyo at present anticipates. At first glance it might appear that in the area of natural gas imports, dependency on the USSR will be substantial in little more than a decade; 24.4 percent of Japan's liquified natural gas (LNG) imports would come from successful Siberian projects. The fact remains, however, that gas itself will play only a very minor role in the mix of energy sources anticipated by MITI's projections. Therefore Soviet gas should account for only two percent of the total fuel sources available to Japan by 1990. Overall, a best case projection on all counts find the USSR supplying just 3 percent of Japan's needs for imported fuels in 1990 and 2.2 percent of the nation's total fuel supply.[39] This is hardly the stuff of which dependency is made.

CONCLUSIONS

The preceding analysis demonstrates the danger of taking too literally Disraeli's famous caveat about "lies, damn lies, and statistics." This is particularly true with regard to the subject of technology transfer—an area in which, as Maurice Mountain of the Office of the Assistant Secretary of Defense for International Security Affairs observes, quoting Mark Twain, "one can get a wholesale return of conjecture from a trifling investment of fact."[40] Examination of Japan's experience with technology imports should alert the reader to the fact that technology bears little resemblance to a precious metal to be hoarded, popular wisdom notwithstanding. Rather, as Joseph Nye of the Kennedy School of Government, Harvard University, has commented, technology is more aptly compared to a fine Rhine wine that

39. Sternheimer, "Japanese-Soviet Energy Cooperation," Final Report, pp. 48-57 (Tables V-VIII).

40. Maurice J. Mountain, "Technology Exports and National Security," in *Issues*, p. 22.

must be drunk when ripe—and shared—if one is not to be left with a cellar full of bottles of vinegar.[41]

We have argued implicitly throughout this article for the need for better understanding—and more precise calculations—of the economic and political as well as military security concerns that cement the diverse interests of Western nations into an alliance. It may at some point be necessary for policymakers in Washington to consider seriously whether or not the advantages to be derived from retarding Soviet economic, technological, or energy development by some three to five years is always worth the price of the ill-will and intraalliance strains that such decisions have entailed in the past. Minimally, it will be necessary to recognize—the thrust of our statistical "snapshots"—that Japanese and Western European interests must invariably suffer in the process. Moreover, the absence of any sympathy for the various kinds of "linkage" arguments advanced by Washington in either Tokyo or European capitals renders the sense of felt deprivation all the more poignant.

From the perspective of the issues specific to the Japanese experience, our analysis underlines the fact that fears of future Soviet competitiveness Japanese-style, even given substantial TT between West and East, are largely unfounded. Likewise, concern over a growing dependency on the USSR among Western allies, as a consequence of repayment for technology trade in raw materials/ fuels, is—at least for the Japanese— without substantial foundation in fact. Japanese-Soviet technology trade is far from critical to Tokyo in

terms of the global distribution of Japanese markets. Nor is Japan the only source of the Western technology that Moscow seeks, even though it is one among several suppliers. The Soviet Union needs Japan far more than Japan needs the Soviet Union, thus limiting Moscow's opportunities for political leverage. Moreover, while Japan's contribution to Soviet energy development in Siberia may be substantial, and here the picture calls for considerably more refinement, the pattern in this area is in no way different from that prevalent in Japanese-Soviet TT considered more broadly.

From a policy perspective, it follows from all that has been said that Washington must proceed more consistently as well as sensitively to avoid giving the appearance to our allies of the boy who cried wolf. Either deliberate consultations and skillful bargaining to build a consensus among the Japanese, West Europeans, and Americans on TT matters or a strong-arm, "damn-the-torpedoes" approach would probably be more effective than the present policy of tacking and veering. Unfortunately, as one observer has put it in a slightly different context, American strategy over the past few years in this area resembles nothing so much as adherence to a dictum of "Speak stickly but carry a big soft."[42] In all probability, such appearances only reinforce Soviet convictions regarding the West's lack of resolve in matters of technology trade. They may also make the aging bureaucrats of the Kremlin more hopeful than ever that Lenin's prophecy will soon be fulfilled.

41. Joseph J. Nye, "Technology Transfer Policies," in *Issues*, p. 16.

42. Major John Hasek of *Canadian Defense Quarterly* at a workshop on "Low Intensity Conflict and the Integrity of the Soviet Bloc," University of New Brunswick, March 23-24, 1981.

Report of the Board of Directors to the Members of the American Academy of Political and Social Science for the Year 1980

MEMBERSHIP

MEMBERSHIP AS OF DECEMBER 31

Year	Number
1970	24,544
1971	23,413
1972	21,963
1973	21,070
1974	19,473
1975	16,923
1976	15,516
1977	14,202
1978	12,816
1979	10,884
1980	10,059

FINANCES

Our bank balance at the end of 1980 was $72,966.39

SIZE OF SECURITIES PORTFOLIO

MARKET VALUE AS OF DECEMBER 31

Year	Value
1970	616,429
1971	612,046
1972	642,808
1973	533,024
1974	371,004
1975	440,450
1976	504,046
1977	451,545
1978	385,795
1979	377,915
1980	368,926

PUBLICATIONS

NUMBER OF VOLUMES OF *THE ANNALS* PRINTED (6 PER YEAR)

Year	Number
1970	145,456
1971	139,450
1972	143,360
1973	132,709
1974	120,397
1975	104,049
1976	101,789
1977	91,367
1978	85,605
1979	71,513
1980	65,153

NUMBER OF VOLUMES OF *THE ANNALS* SOLD (IN ADDITION TO MEMBERSHIPS AND SUBSCRIPTIONS)

Year	Number
1970	14,143
1971	10,046
1972	16,721
1973	12,430
1974	13,153
1975	13,034
1976	12,235
1977	6,296
1978	8,124
1979	5,907
1980	8,751

INCOME STATEMENT FOR THE YEAR ENDED DECEMBER 31, 1980

INCOME:

Dues and Subscriptions	$168,497.05
Sales of Publications	27,415.40
Sales of Review Books	1,667.00
Advertising Revenues	12,054.75
Royalties and Reprint Permissions	5,188.44
List Rentals	2,122.21
Rents	1,300.00
Donations	250.00
Miscellaneous Revenues	789.61
Total Revenues	$219,284.46

OPERATING EXPENSES:
Publications—Printing, Binding & Mailing...$99,744.59
Cost of Publications Sold...5,244.79
Salaries $114,399.48
Payroll Taxes 10,291.02
Pension Expenses—Funded 14,290.40
Pension Expenses—Unfunded 4,181.76
    Total Salaries and Fringe Benefits 143,102.66

Bad Debt Expense...58.20
Depreciation 1,184.00
Insurance...2,412.79
Postage 2,910.01
Printing, Duplicating & Stationery...1,048.95
Promotion...18,796.84
Supplies...3,495.94
Telephone 2,116.73
Miscellaneous Expenses...20,259.86
Repairs and Maintenance 5,715.63
Utilities 4,644.36
    Total Operating Expenses 310,735.35
    Operating Loss ....($91,450.89)

OTHER REVENUE (EXPENSES):
Investment Income...$31,538.01
Less: Investment Fees..1,970.44 $29,567.57
Gain on Sale of Investments 37,424.15
Grant Administration Overhead..17,668.85
    Total Other Revenue 84,660.57
    Loss before Income Taxes...($6,790.32)

Federal Income Taxes (Note 1)..53.87
    Net Loss ....($6,844.19)

Retained Earnings—January 1, 1980 ....304,638.53
    Retained Earnings—December 31, 1980 ....$297,794.34

COMPARATIVE FINANCIAL DATA FOR YEARS SHOWN

	1980	1979	1978	1977
1. CONDENSED OPERATING STATEMENTS:				
Revenues:				
Dues and Subscriptions	$168,497	$195,614	$205,813	$205,818
Sale of Publications	29,082	24,462	38,446	21,866
Advertising Revenues	12,055	7,937	8,352	13,540
Annual Meeting Revenues	—	5,562	7,804	7,732
Miscellaneous Revenues	9,650	10,200	11,628	10,092
	$219,284	$243,775	$272,043	$259,048
Operating Expenses:				
Salaries and Fringe Benefits	$143,103	$153,917	$148,998	$159,401
Publications—Printing, Binding & Mailing	117,824	98,705	100,734	100,191
Other Operating Expenses	49,808	50,639	66,287	51,570
Total Operating Expenses	$310,735	$303,261	316,019	311,162
Operating Loss	($91,451)	($59,486)	($43,976)	($52,114)
Other Revenue (Expenses)	$84,661	$44,456	$34,170	$38,798
Income (Loss) before Taxes	($6,790)	($15,030)	($9,806)	($13,316)
Provision for Federal Income Tax	54	-0-	-0-	1,699
Excess of Expenses over Revenue	($6,844)	($15,030)	($9,806)	($15,015)
2. INVESTMENTS AT END OF YEAR—AT COST	$368,926	$356,451	$355,696	$408,287
3. NET WORTH AT END OF YEAR	$297,794	$304,639	$319,669	$329,476

Report of the Board of Directors

During 1980, the six volumes of THE ANNALS dealt with the following subjects:

January — *The Social Meaning of Death*, edited by Renee C. Fox, Professor of Sociology in the Faculty of Arts and Sciences, in the Departments of Psychiatry and Medicine in the School of Medicine, and in the School of Nursing, University of Pennsylvania, Philadelphia.

March — *The Academic Profession*, edited by Philip G. Altbach, Professor and Chairman, Department of Foundations of Education and Professor of Higher Education, State University of New York, Buffalo; and edited by Sheila Slaughter, Assistant Professor of Education, Virginia Polytechnic and State University, Blacksburg.

May — *New Directions in International Education*, edited by Richard D. Lambert, Professor of Sociology, Chairman of the South Asia Regional Studies Department, University of Pennsylvania, Philadelphia.

July — *Reflections on the Holocaust*, edited by Irene G. Shur, Director, Ethnic Studies Institute, Professor of History, West Chester State College, West Chester, Pennsylvania; edited by Franklin H. Littell, Chairman, National Institute on the Holocaust, Professor of Religion, Temple University, Philadelphia, Pennsylvania; and edited by Marvin E. Wolfgang, President, American Academy of Political and Social Science, Professor of Sociology, University of Pennsylvania, Philadelphia.

September — *Changing Cities: A Challenge to Planning*, edited by Pierre Laconte, Director for University Expansion, University of Louvain, Louvain-la-Neuve, Belgium.

November — *The Police and Violence*, edited by Lawrence W. Sherman, Associate Professor, Graduate School of Criminal Justice, Albany, New York, Director of Research, Police Foundation, Washington, D.C.

The publication program for 1981 includes the following volumes:

January — *America Enters the Eighties: Some Social Indicators*, edited by Conrad Taeuber, Director, Center for Population Research, Georgetown University, Washington, D.C.

March — *America As A Multicultural Society*, edited by Milton M. Gordon, Professor of Sociology, University of Massachusetts, Amherst, Massachusetts.

May — *Gun Control*, edited by Philip J. Cook, Associate Professor of Public Policy Studies and Economics, Duke University, Durham, North Carolina.

July — *Social Effects of Inflation*, edited by Marvin E. Wolfgang, President, American Academy of Political and Social Science, Professor of Sociology, University of Pennsylvania, Philadephia, Pennsylvania.

September — *National Security Policy for the 1980s*, edited by Robert L. Pfaltzgraff, Jr., Professor of International Politics, Fletcher School of Law and Diplomacy; President, Institute for Foreign Policy Analysis, Inc., Cambridge, Massachusetts.

November — *Technology Transfer: New Issues, New Analysis*, edited by Alan W. Heston, Associate Editor, The American Academy of Political and Social Science; edited by Howard Pack, Professor of Economics, Swarthmore College, Swarthmore, Pennsylvania.

During 1980, the Book Department published approximately 300 reviews. The majority of these were written by professors, but reviewers also included university presidents, members of private and university-sponsored organizations, government and public officials, and business professionals. Over 500 books were listed in the Other Books section.

One hundred and twenty-two requests were granted to reprint material from THE ANNALS. These went to professors and other authors for use in books in preparation and to nonprofit organizations for educational purposes.

MEETINGS

The eighty-fourth annual meeting was postponed to 1981.

OFFICERS AND STAFF

Board members Covey T. Oliver, Howard C. Petersen, Lee Benson, and A. Leon Higginbotham, Jr., were reelected for another three-year term. Marvin E. Wolfgang was reelected President for a year and Richard D. Lambert as Vice-President.

The Board also renewed the term of its counsel, Henry W. Sawyer, III, Norman D. Palmer as Secretary, and Howard C. Petersen as Treasurer.

All other officers were reelected and both Editor and Assistant Editor were reappointed.

Respectfully submitted,

THE BOARD OF DIRECTORS

Norman D. Palmer
Howard C. Petersen
Elmer B. Staats
Marvin E. Wolfgang
Lee Benson
A. Leon Higginbotham, Jr.
Richard D. Lambert
Rebecca Jean Brownlee
Covey T. Oliver
Thomas L. Hughes
Matina S. Horner

Philadelphia, Pennsylvania
1 November 1981.

Book Department

INTERNATIONAL RELATIONS AND POLITICS

MICHAEL HANDEL. *Weak States in the International System.* Pp. xv, 318. Totowa, NJ: Frank Cass, 1981. $27.50.

The most interesting feature of this thoroughly researched and thoughtful volume is the clear impression it conveys that we cannot develop many potent generalizations about states simply by categorizing them in terms of relative power via selected indicators. Handel's qualitative approach, drawing on numerous historical examples, challenges many propositions in the literature on small or weak states while his own conclusions are carefully qualified as well. Examples will illustrate the flavor of the book.

Many analysts deem it useful to label certain states as "weak" for purposes of explaining their behavior. But Handel demonstrates that weakness is relative and that many of the criteria used to detect it are imprecise, inadequate, and inaccurate. He finds it necessary to construct an ideal type for weak states (and

another for strong states) because any other description ignores important dimensions and fits too few states. He effectively demonstrates (1) that many weak states can benefit from strengthening themselves militarily and (2) that many can bask under a strong state's protection while neglecting their military strength. Both are correct, and, as they also apply to states that are not weak, describing a state as weak does not tell us much.

One Handel premise is that state behavior is largely determined by power relations and differentials among states. Yet he finds no connection between a state being weak and the degree to which it is nonaggressive. Also, he argues that the behavior of great powers has been inhibited by the spread of liberal-egalitarian-democratic norms in international relations and by nuclear weapons, making power less important. When he argues, convincingly, that weak states have more flexibility and enjoy more success in international economic relationships than is customarily believed, this again qualifies any conclusion about the importance of power.

Another Handel premise is that system structure determines weak states' behavior. But then he stresses that maneuverability is critical in shaping weak states' actions and that maneuverability depends on the dynamics of the international system at any particular time plus the geographical location of the state-making structure and is only a partial guide to weak states' actions.

Thus this is a balanced, judicious review of the complexities and subtleties we must comprehend to understand many nations' activities. In a few places the discussion is out of date, and often the data cited are a decade old. Otherwise the book is clear, well organized, and contains an excellent bibliography. Many people will find it rewarding, if they are not intimidated by the price.

PATRICK M. MORGAN
Washington State University
Pullman

LEON HURWITZ. *Contemporary Perspectives on European Integration: Attitudes, Nongovernmental Behavior, and Collective Decision Making.* Pp. xx, 292. Westport, CT: Greenwood Press, 1980. $27.50.

Leon Hurwitz, an American political scientist, edited this collection of 12 scholarly essays that week to ascertain how far the European community has promoted true integration of Europe's economic, social, and political life. Several of the contributors are American and European political scientists; the others are specialist in law, communications, economics, and medicine. Most have had direct experience with the EC, and some are past or present community officials.

Hurwitz groups the essays into three parts that reflect corresponding perceptual approaches to the study of European integration: integration at the individual's (or groups of individuals') attitudinal level; integration at the societal level; and integration at the level of formal governmental behavior

and collective decision making. Part I contains two essays. Ronald Inglehart, an American political scientist, and Jacques-Rene Rabier, a former EC Director-General, are authors of "Europe Elects a Parliament: Cognitive Mobilization, Political Mobilization, and Pro-European Attitudes as Influences on Voter Turnout." They reject the notion that the June 1979 European Parliament elections saw a swing to the right across Europe. Only in Britain was this the case. Elsewhere the tendency was for the relatively pro-European parties to gain ground. In "Phantom Europe: Thirty Years of Survey Research on German Attitudes toward European Integration," Elisabeth Noelle-Neumann of Germany finds that although her fellow citizens were prepared after the war to accept a "supranational" structure, their confidence in this has been shaken since 1974—the reasons being their fear of inflation at home and their fear of increasing Communist influence within the European Parliament.

Part II concerns European integration at the societal level. In "Trade, Interdependence, and European Integration," Wilfried Prewo, a German economist, undertakes a transactional flow analysis. He believes that further reductions in nontariff and language-related barriers will provide a rich source for future integration effects. In "Interest Group Behavior at the Community Level," Emil Kirchner, an English political scientist, predicts that trade unions and social, consumer, and environmental interest groups will exert a stronger voice at the community level.

Part III, which approaches European integration at the level of collective governmental decision making within the community, contains five papers. In "Two-Tier Policy Making in the EC: The Common Agricultural Policy," Werner Feld, an American political scientist, examines the CAP in detail. He thinks the CAP is likely to survive, although its basic nature will probably be much different than originally conceived, especially with the entry of Greece, Portugal, and Spain into the EC. In "The Delicate Balance Between Municipal Law and Community Law in the Application of Articles 85 and 86 of the Treaty of Rome," Stephen Kon of England studies the leading decisions of several national courts and the European Court of Justice. He finds that the problems generated by this overlapping of legal systems have been resolved by making the commission and "European" law prevail over their respective national counterparts. In "Health Policies in the EC: Attempts at Harmonization," two French doctors, Roger Vaissiere and Jean-Marc Mascaro, study how the 1975 decision to permit "free circulation" of EC physicians is actually working. Thus far relatively few have taken advantage of the decision. Robert J. Lieber's essay, "Energy, Political Economy, and the Western European Left: Problems of Constraint," discusses the international energy crisis and the EC's energy policy (or lack thereof) in terms of domestic European political realities for both existing governments and possible future ones. In "European Fiscal Harmonization: Politics During the Dutch Interlude," Donald J. Puchala of Yale presents a case study of Dutch behavior during 1964-67 when the EC's proposed value-added tax (VAT) was almost brought to a standstill.

Part IV deals with decision making, decisions, and policies that relate to countries other than the EC members. There are two essays: "Foreign Policy Formation Within the EC with Special Regard to the Developing Countries," by Corrado Pirzio-Biroli, an Italian developmental economist; and "EC-U.S. Relations in the Post-Kissinger Era," by Harold Johnson, an American political scientist. The latter will be of particular interest to American readers. Johnson points out that not until 1977 was the EC directly represented in the discussions among the "Big Seven" industrialized countries of the free world. He concludes that the EC will require a more integrated foreign economic policy before

it will be able to negotiate with the United States as a unit and be seen as a truly equal partner.

CHARLES F. DELZELL
Vanderbilt University
Nashville
Tennessee

LAWRENCE S. KAPLAN. *A Community of Interests: NATO and the Military Assistance Program, 1948-1951.* Pp. xii, 251. Washington, DC: Office of the Secretary of Defense, 1980. No price.

The North Atlantic Treaty (as the author of this small monograph noted half a dozen years ago) is a subject which historians have for the most part been content to leave to the political scientists. But in his study of U.S. military assistance after 1945, which is a revised version of a work begun in the early 1950s, Kaplan has contributed an important chapter to the story of how paper commitments and plans were given substance. His account, which begins with the Truman Doctrine and is centered on, though by no means confined to, assistance to Europe, brings out clearly the problems involved: difficulties at home—interagency rivalries and personal differences within the administration; dealing with Congress and public opinions—and abroad; the sensitivity of the Europeans; anxiety (which was shared by many on the United States side where the arguments for economic aid found a warmer reception in Congress than did military aid) not to compromise economic recovery by diverting resources to military buildup; and above all, the key question which came to the fore in that troubled summer of 1950— how to strengthen Europe without involving some military commitment from the Federal Republic of Germany. It is a complex tale, especially on the administrative side, as it involves a dissection of a multiplicity of agencies with their confusing titles and initials which wrestled with the problems on both sides of the Atlantic. The Military Assistance

Program, Kaplan concludes, "played a vital catalytic role in mobilizing the forces of the United States and Europe to defend the Alliance." In this process the Korean war is a landmark; but the evidence produced suggests that it was the chance confluence of the crisis in the Far East with the puzzling over how to match Soviet strength in Europe as reflected in the National Security Council's report of April 1950 rather than any rational or predetermined course which decided United States policy—a conclusion which, as Kaplan notes, will provide little comfort to revisionists. Paradoxically, too, once the initial anxiety passed, the Korean war failed to jar NATO from its lethargy, even though by the end of 1950 it had emerged as a genuine military organization. Kaplan's account of how "the Military Assistance Program served to energize the Alliance without destroying political controls over military policies and without debasing the Allies to the position of satellites" is a welcome addition to the literature on the period and is especially valuable for the solid way in which it is based on unpublished sources in the files of the Office of the Secretary of Defense and elsewhere.

ROBERT SPENCER

University of Toronto

STEIN UGELVIK LARSEN et al., eds. *Who Were the Fascists? Social Roots of European Fascism.* Bergen: Universitetsforlaget, 1980 (Distributed by Columbia University Press). Pp. 816. $48.00.

This massive volume, carefully edited by Larsen and seven colleagues, is the product of a conference held in Bergen, Norway, in June 1974. The editors then solicited additional manuscripts in order to present a truly complete survey of European fascist movements, and arranged for excellent translations of the papers not originally delivered in English. There are 44 contributors, teaching in the universities of 17 different nations, and a total of 51 papers

including introductions and a preface. The principal scholars working in the field are in large measure represented. Some eminent specialists, such as George Mosse and Barrington Moore, are not among the contributors, but their writings are extensively discussed (11 references for Mosse, 16 for Moore).

The work is divided into six sections, with a focus on areas where fascist movements are not as well known. (For example, the section on the "Fascist Core Countries: Germany and Italy" occupies only 91 pages; while the treatment of "Fascism and National Socialism in the Nordic Countries," where percentages of national socialist party members were extremely low—even in occupied Denmark and Norway—covers 165 pages.)

In a brief review it is impossible to do justice to such a complex and diverse work. *Who Were the Fascists?* certainly belongs in college and university libraries, and political scientists will find it useful for courses in comparative totalitarian systems, and doubtless for many other purposes. It is a veritable mine of information, making extensive use of tables and electoral statistics. Where else can one find gathered in one volume a discussion of the tiny national socialist movement in Iceland (which continued to hold cell meetings until 1944, when it became clear that Germany was losing the war), an elegant paper on Swedish fascism, which never attracted more than 0.14% of the electorate, thoughtful analyses of fascism in such countries as Romania and Switzerland, and a brilliant paper by Philippe Schmitter on the Portugal of Salazar?

The approach is resolutely interdisciplinary, and the papers range from highly abstract and occasionally fanciful speculation to precise and systematic deconstruction of older theories. As Stanley Payne remarks in an excellent essay introducing the first section, "writing the history of fascism today inevitably means writing the history of scholarship on fascism."

For all the sophisticated analyses found in so many of these papers, sometimes a simple and straightforward explanation seems most persuasive. Bruce Pauley concludes his treatment of the struggle between Nazis and native Austrian fascists by observing that in the end the Nazis were able to absorb all the authochthonous fascists because of the *Anschluss* of March 1938. "Success was the absolute *sine qua non* for fascists not only in Austria, but also elsewhere in Europe. Nothing else could hold their heterogeneous social groups together." This statement might have served as a n epigraph for the entire volume.

DAVID L. SCHALK
Vassar College
Poughkeepsie
New York

ARCHIMEDES L.A. PATTI. *Why Viet Nam?: Prelude to America's Albatross.* Pp. xx, 612. Berkeley: University of California Press, 1980. $19.50.

PAUL M. KATTENBURG. *The Vietnam Trauma in American Foreign Policy 1945-75.* Pp. xvi, 354. New Brunswick, NJ: Transaction Books, 1980. $19.95.

WILLIAM S. TURLEY, ed. *Vietnamese Communism in Comparative Perspective.* Pp. xiii, 271. Boulder, CO: Westview Press, 1980. $22.50.

There is an awesome timeliness about these three new books on Vietnam, which were published coincidentally as one controversial administration was drawing to a close and another equally controversial one was preparing to begin. The context in which these books—in no respect are they overlapping—appear strongly suggests that at least some of the lessons of Vietnam have been learned, however imperfectly, by the American people, as well as by the international community. It now seems unlikely that a willful executive can have his way without a most serious public challenge to his determination.

Why Viet Nam? was written by the man who commanded the OSS detachment in Tonkin and probably knew Ho

Chi-minh better and more intimately than any American who has ever lived. Although rigidly and sometimes frustratingly circumscribed by his instruction to maintain American neutrality in the Franco-Vietnamese conflict, Patti and Ho came to like and respect each other, even to reach a degree of mutual affection. It was Patti who transmitted Ho's appeals to President Truman for some kind of recognition or at least some gesture of understanding—appeals which were never even acknowledged as Washington in the post-Roosevelt period moved steadily into a more pro-French and anti-Communist stance.

The end of the war against Japan in the fall of 1945 was probably the last chance to halt, or even moderate, an American anti-Communist policy which still insensitively haunts our attitudes. And yet, Patti was convinced that Ho, who never in any way denied his Marxism, was first and foremost a Vietnamese nationalist, and only then a Communist. It was a subtle distinction on which men could disagree.

Paul Kattenburg, a twenty-year veteran of the American Foreign Service who worked primarily on Vietnam, picks up the story where Patti leaves off. He does not, however, limit himself only to Vietnam, but places that benighted country in the broader framework of overall American foreign policy. In my view, he succeeds magnificently.

Space precludes going into detail on how the men around Kennedy and Johnson, men of unquestioned ability, the "best and the brightest," could have locked themselves into the positions they did. Basically, it was ghastly bad political judgment. The question of morality really does not arise until Nixon and Kissinger abrogated all decisions to themselves, "peace with honor," and then deliberately dragged on the war for another long four years until Nixon had won his unprecedented victory in 1972.

A couple of points do warrant special mention. Vietnam was a war in which all decisions were made by civilians. Technically, General Westmoreland is quite right that the war was not lost on the battlefield. The cold fact is that the Joint Chiefs were never fully consulted about anything; seldom was their advice even sought. The war was lost at home when the public began to realize it was being lied to, that the war was winnable only at a cost the United States was in no sense prepared to pay and, finally, that the core of South Vietnam was just as rotten and unworthy as Chiang Kai-shek's China had been, as North Vietnam would so devastatingly demonstrate somewhat to its own surprise in 1975. Clemenceau's famous comment about generals was being carried to self-defeating extremes.

A second point is that no administration ever seriously thought it could win a war the French had lost, as De Gaulle rather irritatingly kept pointing out to Washington. (France, incidentally, also did not lose in Vietnam; it lost at home for reasons not dissimilar to the American ones). Washington simply could not bring itself to admit it had lost a war to a Communist state, except on battlefield terms that never seemed propitious until Johnson finally quit in the spring of 1968. Nixon then raised the negotiating ante to levels he probably knew North Vietnam saw no reason to accept. It was public reaction to Cambodia and the Christmas bombing of Hanoi that finally persuaded Nixon to accept in 1973 terms Johnson could probably have had just as easily in 1968. Vietnamization was a fraud, and everyone except President Thieu finally knew it. All Vietnamization accomplished was to make a united Vietnam a major military power after 1975. If North Vietnam correctly assessed American attitudes, as it did, it is less certain that it understood enough about American politics to have foreseen Watergate as the inevitable last chapter in a Greek tragedy.

Vietnamese Communism in Comparative Perspective does something which, to my knowledge, has not been done before. The Vietnamese Studies Group, a committee of the Southeast Asia Council of the Association for Asian Studies, organized the seminar in 1978 which produced the papers in this volume, all written by a group of younger French and American scholars. The papers are uniformly good, each with something special to say. Although no categorical judgments are made, what is at least implicit in all the papers is the distinctiveness of the Vietnamese experience, finally concluding that it is a focal point in the socialist world.

Clearly, there have been ups and downs in the Vietnamese relationship with the Soviet Union, depending in large measure on the vagaries of the Moscow attitude toward Hanoi. Vietnam has had a lot of experience at playing off one major power against another.

It is the relationship with China that is the most illuminating. There was a Chinese military staff advising General Giap at Dien Bien Phu. At the last moment he chose to ignore their advice and the seige was delayed for two months. Similarly, the land reform movement of 1955-58 started out according to the lines of Maoist populism and millenarianism. Within a very few months it became apparent this was not suitable for Vietnam and appropriate changes were made before too much damage had been done. As for Cambodia, Hanoi was distressed to learn how fascinated Pol Pot had become with his reading of Maoism and agriculture. When Pol Pot came to power, Hanoi was appalled, as was the world, by his excesses and invaded Cambodia as well as strengthening its ties with Moscow. Peking's response came quickly "to punish" Hanoi. The cost to both sides was dreadful, but Hanoi did discover that PLA tactics and strategy had not moved much beyond the days of The Long March. Even some of the Chinese generals were aging hangers-on from that epic period.

One of the real tragedies of the Vietnam trauma, as Kattenburg points out, was that we have lost or forgotten the art of diplomacy. It is fashionable now to sneer at the diplomacy of the nineteenth century, but with local exceptions it did give us a full century of peace, a century filled with new and exciting ideas, a century to consolidate—at a cost, to be sure: the industrial revolution. War then, with all its miseries, was an acceptable alternative. In the nuclear age, it no longer is. Henry Kissinger, whatever his faults, understood the uses of diplomacy, the art of compromising the irreconcilables, of learning to live with ways one finds distasteful. Whether we can learn or relearn this is still to be seen.

JOHN F. MELBY

University of Guelph
Ontario
Canada

AFRICA, ASIA, AND LATIN AMERICA

RICHARD B. BARNETT. *North India Between Empires: Awadh, the Mughals, and the British, 1720-1801.* Pp. xviii, 294. Berkeley: University of California Press, 1980. $25.00.

During the eighteenth century, perhaps the most politically complex epoch in Indian history, the Mughal Empire disintegrated and the English East India Company became an Asiatic power. Though these developments have naturally attracted research and spurred debate from the very outset of British rule, only in the last two decades have historians probed deeply the Indian regional polities that emerged from the crumbling Mughal domain and confronted expanding European dominance. Richard Barnett's book is the first monograph in this recent line of scholarship, and the polity in question is particularly important. Awadh was not only the largest, wealthiest North Indian regional kingdom, it was also the closest to Delhi, the Mughal capital, and its final annexation by the company in 1856 helped trigger the great Indian uprising of 1857.

Barnett begins with a useful sketch of current research on Mughal decline and regional state-building. He goes on to dissect the rise of Awadh as an autonomous domain, its increasingly subordinated relation with the company, its faltering status as a North Indian subsidiary power, its spasmodic blows against company domination, and, finally, its partial dismemberment by company annexation in 1801. Drawing on a variety of native language sources, and building from a rich body of secon-

dary literature, Barnett concentrates on the personalities and human interactions at the heart of the Awadh political system. He shows us both a system at work and real people doing the best they can in the midst of dramatic historical transformations.

As the title suggests, the book concerns the North Indian political system, as seen through the ruling elite in Awadh, rather than the Awadh political system itself. Barnett is not concerned here with the structure of the system, or with links between elite behavior and conditions among Awadh's millions of peasants. This, therefore, is a study of high-level political strategies, motivations, and struggles; as such, it adds new richness to our understanding of early modern elite political behavior at a critical point in Indian history. It does not, however, explain the evolution of the political system as a whole, or the structural course of political fragmentation and integration in early modern India.

DAVID LUDDEN

University of Pennsylvania
Philadelphia

YAACOV BAR-SIMAN-TOV. *The Israeli-Egyptian War of Attrition, 1969-1970: A Case Study of Limited Local War.* Pp. xi, 248. New York: Columbia University Press, 1980. No price.

From March 1969 to August 1970, the Egyptians and Israelis fought one of the most peculiar wars in their long history of conflict. Called the War of Attrition, it was limited in character but bloody in effect. According to Bar-Siman-Tov, Egyptian military casualties sometimes reached as many as 300 a day. Official Israeli figures, perhaps overly sanguine, admit to 341 such casualties from October 1969 to July 1970. Officially, both sides claimed to have won the war, in both military and political terms. The author, on the other hand, tends to think both lost: "Unlike the wars that preceded it, the War of Attrition ended without a clear-cut military decision. It ended in a strategic draw and the wearing down of both contestants." But it also ended through the intervention of the United States and the Soviet Union. What, then, is there to make of this nasty, fruitless, and embittering episode?

Certainly there are many ways to approach this question. Unfortunately, Bar-Siman-Tov has chosen to study it from a theoretical, almost clinical perspective, in which the War of Attrition is treated essentially as a case study that may illustrate something about limited wars more generally. Thus, the first chapter is riddled with such concerns as definitions of limited wars, constraints and limitations upon them, bargaining in such situations, escalation, dynamics, war management, and patterns of termination. This approach creates three major problems.

First, the single case study sheds little light on the theoretical framework, and the theoretical framework tends to obfuscate the case study. Indeed, Bar-Siman-Tov's last paragraph is an admission of the limitations of the study's design: "Additional case studies of limited local war are needed, of course, in order to broaden understanding of the range and variety of interrelationships between sources of constraints and limitations in local war, their interactions, and their roles in determining the outcome of such conflicts."

Second, the sources of information available to the author were so unbalanced that one side of the case study is inherently speculative Thus, Bar-Siman-Tov admits, or claims: "Being an Israeli I could not interview any Egyptian officials and the study lacks this kind of information." How, then, does he come to start sentences with "the Egyptians thought" or talk about "the war aims of the parties"? Relying upon public information from the Egyptian side, much of it propagandistic and questionable, is not suitable to a study of this kind. Apparently the author must recognize this as he maintains that on the Israeli side perceptions of the war differed according to whether one was within the government or outside of it.

Third, despite some lip service to past and subsequent wars and to domestic politics, the book is largely ahistorical and apolitical. For instance, what impact did the War of Attrition have on the continuing Middle-Eastern conflict, and what impact did it have on the political fortunes of those politicians and military officials who commanded it? These are questions that could have been, but were not, answered in this study.

Had Bar-Siman-Tov's single case study somehow confirmed a grand theory of limited local war, or had it treated the War of Attrition on its own historical and political terms, it would have been extremely valuable. Unfortunately, accomplishing neither, the book loses much of its potential value.

DAVID H. ROSENBLOOM
Syracuse University
New York

JOEL S. MIGDAL. *Palestinian Society and Politics.* Pp. xv, 292. Princeton, NJ: Princeton University Press. 1980. $20.00. Paperbound, $6.95.

SIMHA FLAPAN. *Zionism and the Palestinians.* Pp. 361. New York: Barnes and Noble Books. 1979. $24.95.

These two recent books approach the Palestinians from two quite different perspectives: Migdal and his collaborators examine them essentially from within, while Flapan deals with them from without.

Palestinian Society and Politics is actually two books in a single volume. Book I, by Migdal himself, is a monograph that attempts to provide an overview of the major transformations undergone by Palestinian society and politics over the last hundred years. Migdal offers a well-reasoned, systematic analysis of Palestinian social and political development under Ottoman, British, Jordanian, and, most recently, Israeli rule. On the broader theoretical level, Migdal tries to uncover some of the main causes and conditions that deny a society social cohesion and coherent patterns of stratification. He develops a model that groups the policies of the ruling regime into three broad categories: (1) investment policies, (2) political alliances, and (3) security policies. In tracing the policies of the four regimes under which Palestinians have lived, he concludes that the policies leading to Palestinian social fragmentation began with the Ottoman reform of traditional land tenure, that the pattern established was then intensified and institutionalized by the Mandatory authorities, and that the investment policies of both the Hashemite Abdallah and later the Israeli government only accentuated the fragmentation. Migdal's rather condensed, almost sketchy style is bolstered by a number of useful tables and graphs.

Book II consists of eight essays by different authors, each examining in some detail a historical or contemporary aspect of the Palestinian social or political structure. All of the essays are well done and on the whole eschew the unfelicitous jargonese of Book I. The excellent contribution, "The Office and Function of the Village Mukhtar," by Gabriel Baer makes available for a wider English-speaking audience an extract from his important Hebrew study on that subject. Particularly noteworthy also are Mark Heller's "Politics and Social Change in the West Bank since 1967" and Kennety Stein's "Legal Protection and Circumvention of Rights for Cultivators in Mandatory Palestine."

In the second book under review here, Simha Flapan, editor of the well-known monthly *New Outlook* and director of the Israeli Peace Research Society, deals with the Palestinians not qua Palestinians, but as the objects of Zionist policy from the early days of the British Mandate to the Israeli War of Independence. Flapan seeks out the roots of contemporary Israeli policy toward the Palestinian Arabs as well as Israeli attitudes toward Palestinian aspirations to nationhood. He argues with some cogency that these policies and attitudes were molded in the prestate period. His point of departure from such writers as Walter Laqueur and Michael Brecher is that all of the important Zionist leaders shared the same misconceptions. He implies that none—not Weizmann, not Ben Gurion, not Sharett—really accorded much

priority to peace with the Arabs—none, that is, except Nahum Goldmann. In Flapan's view "Goldmann was the only Zionist leader to grasp the dynamics of Arab nationalism." He appears throughout the book as a great visionary and consummate statesman. It should be noted in passing that Goldmann is thanked in Flapan's acknowledgements for having provided the funds for researching the book.

Though interesting and eminently readable, *Zionism and the Palestinians* is not very satisfying. Flapan rejects the view of most historians of Zionism that the Jewish Agency in Mandatory times, and later the Israeli government, never had any real alternatives for action open to it, but rather faced "real choices and *was aware of it.*" However, he never really spells out what those choices were and even admits that "the question of whether an alternative orientation was possible for the Zionist movement begs many questions." This is not a scholarly book and makes no pretensions to be one. It relies in the main on secondary works and published records, although there is considerable use of archival material from Jerusalem, London, and Washington. Flapan provides few answers, but he raises many thought-provoking questions.

NORMAN A. STILLMAN
State University of New York
Binghamton

ROBERT A. MORTIMER. *The Third World Coalition in International Politics.* Pp. xii, 148. New York: Praeger Publishers, 1980. $16.00.

ELLEN FREY-WOUTERS. *The European Community and the Third World: The Lome Convention and its Impact.* Pp. xii, 291. New York: Praeger Publishers, 1980. $22.00.

These two books complement each other neatly; their substantive discussions are intertwined, but each has a distinctive focus in the treatment of a common theme. The two authors address virtually the same issue, but they approach it from two different perspectives so that the topics they choose to explore are related and yet distinct. The first basically recounts the tenacious struggle among Third World nations to organize themselves for a meaningful role in international politics. The second presents an analysis of a significant conclave organized to create a cooperative relationship between the developing and developed nations. Both studies are concerned with the changing nature of the political and economic relationships between the powerless, the poverty-stricken, the former colonial possessions in Africa, Asia, and Latin America on the one hand, and the affluent, the powerful, the former colonial powers of the West on the other. Both illuminate the reader with massive details on the behavior of nations in the broad context of world affairs, though the focus of one is narrower and deeper than the other.

The work by Mortimer is essentially a description of the efforts by the countries of Africa, Asia, and Latin America to forge a coalition in order to become an effective force in international politics. In this regard Mortimer presents the larger picture; Frey-Wouters's analysis constitutes a subset in that domain. Mortimer examines the evolution of a common Third World consciousness and submits that the helplessness of the newly independent nations in the face of the superpower domination of the international system sharpened their search for a forum to articulate common concerns. The attempt to coalesce their interests began in the celebrated Bandung Conference of 1955 with continuing series of conferences following, some to devise strategies for attainment of power, others "to protest against the subordinate status of the developing countries in the international system" as well as efforts by various leaders to seek "to materialize the spirit of Bandung in more concrete terms."

Despite enormous diversity among the countries of the Third World, the Bandung experience created an unprecedented sentiment for solidarity; and, according to Mortimer, two of its goals—decolonization and economic development—remain the most articu-

lated for achievement to this day. Although the Third World coalition first demonstrated its power at the Sixth Special Session of the United Nations General Assembly in April-May 1974 by demanding the creation of, among other things, a new international economic order, Mortimer is quite realistic in suggesting that the "Third World power is a fragile construct, depending as it does upon a collective cohesion that is difficult to maintain." He perceptively argues, nonetheless, that the Third World mobilization of power is a fact of life that ought to be accorded legitimacy, and that, rather than maneuvering to divide and weaken the coalition, it should be recognized in the interest of greater global equity.

In this straightforward narrative account Mortimer has produced an interesting study that fills the gap in our knowledge of the Third World's contribution to international politics. It is a piece of scholarly research that is neither contentious nor theoretical. Consisting primarily of factual description, the book is sensitively put together, capturing the aspirations of a powerless people much maligned during the last two centuries.

Frey-Wouters arrives at virtually the same conclusion as does Mortimer, although in her estimation Third World aspirations are to be fulfilled through economic action conducted in a spirit of collaboration, rather than adversarial political action. For Frey-Wouters, the development of economic bonds in which power is shared is a step in the right direction; consequently, her work is devoted exclusively to an analysis of a major forum for economic negotiations: the Lome Convention. She examines the events that led up to it, assesses its impact on the world community, and projects its meaning for the future. The author uses this conference as a vehicle for pointing out that the development of cooperative relationships between potential antagonists is possible, even though it may involve difficult and sometimes cumbersome negotiations. The Lome Convention was the culmination of talks initiated in the early

1960s to find means for cooperative relationships between some of the members of the European Economic Community and several countries of Africa, the Caribbean, and the Pacific region. Frey-Wouters looks at the entire process in great detail and finds that the Lome Convention has come closest to establishing a new pattern of trade and economic relationships between the developing and developed countries. Although the Lome Convention is a marked improvement in the unequal ties that existed between the colonizer and the colonized, there is still considerable distance to be traveled before a just and equitable economic order can be established.

Frey-Wouters presents a staggering amount of data on a variety of resources and economic variables, and uses these as a basis for making her arguments that a fundamental change in the existing structures of economic relationship is not only desirable but mutually beneficial for the developed and developing nations. She believes that the resources of the world can be shared on the basis of justice and equality. Partnership with the have-nots is "the only option before the Western world" if impending economic disasters are to be averted.

This is a thoroughly researched, well-documented, analytical study. It is a substantial work in which the author demonstrates an extraordinary command of the knowledge of the subject under discussion. She is quite bold, too, mincing no words in making her policy recommendations, which seem reasonable enough but would be disturbing to those in the corridors of power.

The two books considered here are suffused with factual details that vividly describe the world undergoing change. The authors untangle a web of complex relationships to make confusing processes easily understandable. In spite of the sterile nature of the subject matter, the books are written with compassion and sensitivity.

GHULAM M. HANIFF
St. Cloud State University
Minnesota

ELIZABETH J. PERRY. *Rebels and Revolutionaries in North China 1845-1945* Pp. xvi, 324. Stanford, CA: Stanford University Press, 1980. $25.00.

In *Rebels and Revolutionaries in North China 1845-1945*, Elizabeth J. Perry has presented an anthropological study of some of the underlying causes of peasant revolt in China. She has taken as the specific area of analysis the Huai-pei region, which is in eastern China lying between the Huai River on the south and the Yellow River on the north. Huai-pei is noted for peasant revolts and for its harsh ecological history and marginal farming conditions. Perry presents evidence that such conditions as infertile land subject to frequent flooding and deposits of sterile sand and pebbles, extensive waterlogging, periodic plagues of locusts, high rents and loan interests, and the hardship of too-small farm plots caused by the Chinese multipartite system of land inheritance forced peasants to adopt survival techniques that included both predatory and protective strategies. These strategies then led to various forms of rebellion and revolution.

Perry began her investigations on the theory that Chinese rebellions and revolutions were inspired by secret societies such as the White Lotus Sect. Her extensive research into primary sources in Taiwan and Japan (the files of the Bureaus of Investigation and Statistics, Ch'ing dynasty archives at the National Palace Museum, Kuomintang archives, and records of Toyo Bunka Kenkyujo and Toyo Bunko), however, cast suspicion on religious origins of revolts and insurrections and pointed more toward banditry and extortion. Such a finding should not have come as a surprise, since official accounts frequently blame rebellion on bandits and "bad elements" and deny the existence or justification of organized insurrection.

Perry lays a foundation for her thesis by describing, first, the Huai-pei environment. This is followed by an examination of strategies for peasant survival: supplementing income by other work both within and outside the Huai-pei area; defense and security measures; and/or smuggling, kidnapping, and banditry.

The body of the study consists of an analysis of three insurgency situations: the Nien Rebellion (1853-1868), conducted by widespread but loosely organized bandit gangs; the Red Speer Society, local defense units associated by quasi-religious rites and beliefs; and the Communist activities during and after the Japanese War.

In each of these revolts against existing authority, the root cause is shown to be a struggle for survival and that the peasant participants were motivated neither by lofty spiritual ideals nor by ideological sentiments. In *Rebels and Revolutionaries in North China 1845-1945*, Perry has made a good case.

JAMES D. JORDAN

Alexandria
Virginia

EUROPE

MARQUIS CHILDS. *Sweden: The Middle Way on Trial.* Pp. x, 179. New Haven, CT: Yale University Press, 1980. $12.95.

FRANCIS CASTLES. *The Social Democratic Image of Society.* Pp. xiv, 162. Boston, MA: Routledge & Kegan Paul, 1978. $15.50.

These two books seek to explain something of the politics of Scandinavia. Childs's book is a reappraisal of Sweden's "middle way" (between capitalism and communism), whose unique features he first extolled in a book in the mid-1930s. Castles's book examines the bases of popular support of social democracy in the three countries of Scandinavia and tries to throw light on the origin, program, and way to power of that political movement. Together these books provide a glimpse of the nature and accomplishments of a somewhat unique political phenomenon.

Working with voting statistics, Castles emphasizes the dominant position enjoyed by socialist democratic parties in the national legislatures of Sweden, Norway, and Denmark and the considerably inferior strength of other parties. The socialist democratic parties built on the trade union movements of the respective countries which (as in Great Britain) subscribed to gradualist principles of social and economic changes. Being in the forefront of struggles for popular rights and espousing public reform (but not ownership) through class collaboration were the avenues to power; increasing disposable incomes and steadily pursuing social and economic amelioration proved to be the means of staying in power in each nation, together with a commitment to strong internal party discipline and working-class agendas. Between the depression and the end of World War II these arrangements worked rather well. Democratic Socialism prospered in spite of ruling-class oposition; social stratification proved not to be impervious to change.

Castles devotes a lot of attention to literature that tries to explain the reasons for social democratic successes. Ideology is found to be of minimal importance; the nature and extent of welfare state reforms, within the market mechanism of a capitalist economic framework, are deemed very important. A viable, if weak, political party of the right and a prevalent middle-class social system in each country are identified as primary factors responsible for bracketing the extent of those reforms, especially in health and welfare matters. Income redistribution could be pushed far without seriously weakening economic efficiency and individual freedom. Such redistribution, indeed, went as far as it did mainly because it was compatible with that continued efficiency. It was preferable (to capitalist owners) to direct assaults on property rights.

Castles is at his best when he seeks to explain why this political scheme has prevailed more notably in Scandinavia. Ethnic and cultural homogeneity (facilitating class cooperation), as well as some of the factors noted previously, are cited as reasons.

How well is this politicoeconomic system bearing up over time? This is what Childs has endeavored to ascertain by reexamining the Swedish experience after four decades. He found that the advantages of Sweden's welfare system today are concrete: cradle-to-grave benefits, a general family allowance for children under 16 of $400 a year, maternity (and paternity) payments, holidays for housewives, the most comprehensive health insurance plan in the world, a minimum of five weeks of paid vacation, workers' pensions of up to 83 percent of salary, and more. But the costs of such benefits are staggering, Childs reports. Taxes have increased to confiscatory levels, and business leaders worry about dwindling funds for investment. Inflation and worldwide economic and financial crises have compounded the problem, as have increasing pressures by labor for a larger share in management decisions.

A new national dilemma looms—should Sweden continue developing nuclear energy? The issue brought one government into power and then caused its downfall. Childs shows that Three Mile Island had a devastating impact on Sweden, desperately in need of new sources of energy and confident until then that it had devised safe nuclear reactors and a safe way to dispose of waste. Deeply divided over the future of nuclear energy, the Swedes held a national referendum to decide the issue on 23 March 1980; the vote was affirmative, but anxieties remain. Childs does not believe that Sweden's welfare system is in danger of imminent demise. What has been built up over nearly five decades is not likely to be abruptly undone. But for those who look to Sweden as a model for distributing the better life to a greater number of people, the situation clearly bears watching. The direction to be taken in the immediate future will say a great deal about the resolution of the dilemma of a social system that has taken such long strides down the road of welfareism.

HARRY W. REYNOLDS, Jr.
University of Nebraska
Omaha

RICHARD A. GABRIEL. *The New Red Legions: An Attitudinal Portrait of the Soviet Soldier.* Pp. 264. Westport, CT: Greenwood Press, 1980. $22.50.

RICHARD A. GABRIEL. *The New Red Legions: A Survey Data Source Book.* Pp. 270. Westport, CT: Greenwood Press, 1980. $40.00.

These companion volumes are listed to retail at $22.50 for the first listed above, and $40.00 for the second—or, both volumes as a "set" for $49.94. Perhaps this is an odd way to open a scholarly review of a set of books except to emphasize the fact that the high cost of these volumes may mean that they are being priced out of the reach of a public which should be aware of this extremely important research by Gabriel. The volumes should be required reading for every military planner, budgeteer, politician, and concerned citizen.

Undoubtedly, social scientists will raise questions about methodology, technique, and interpretation of the data—all the while, it is hoped, realizing the significance of what Gabriel has done here. Utilizing the resource of Jewish emigres from the Soviet Union to the United States under recently liberalized emigration policies, he has solicited from 161 of these persons their responses to a lengthy questionnaire regarding their experiences in Soviet military service. The respondents represent officer, noncommissioned, and conscript ranks of all of the Soviet services. The *Survey Data Source Book* includes all responses—computerized, of course. The companion volume contains Gabriel's interpretation of the data.

Together, the books carry an impact reaching far beyond the modesty of the sample size, or quarrels about the suitability of the questionnaire, or, for that matter, differences about interpretations of the data.

Put too simply perhaps, it appears that the Soviet conscript army is not necessarily superior to the Western "volunteer" armies. It suffers from serious leadership problems, low morale, and a lack of peacetime or combat cohesion. It is not ideologically or patriotically motivated despite all efforts of the Soviet military to develop these traits. It, at least, is not convinced of its invulnerability or superiority. The Soviet forces suffer from high AWOL rates, significantly high numbers of suicide attempts, heavy drinking, and a general lack of confidence in the military system. These poorly paid, overworked, severely regimented and harshly punished soldiers are not in the hands of traditional professionally motivated military commanders, but rather under the command of Soviet bureaucrats in uniform—with all that this portends for the existence of a modern combat "able" fighting military force.

Maybe we do not want to know this. Or, again, for comparison's sake—and for control—Gabriel should randomly select 161 U.S. forces veterans to whom the same questionnaire should be submitted and analyzed.

But for opening this Pandora's Box, Gabriel has earned my respect.

JACK L. CROSS

Texas A&M University
College Station

PATRICIA JALLAND. *The Liberals and Ireland: The Ulster Question in British Politics to 1914.* Pp. 303. New York: St. Martin's Press, 1980. $27.50.

The decline of the once powerful Liberal Party has long fascinated British historians. In 1935 George Dangerfield's seminal *Strange Death of Liberal England* commenced the historiographic controversy. According to Dangerfield, the Liberals, already enervated in 1909 by their inability to understand the new working class, soon succumbed to a series of crises spawned by the emerging Labour Party, vociferous suffragettes, industrial conflict, a constitutional confrontation involving the House of Lords, and the Irish question. Several scholars, including Peter Clarke and Trevor Wilson, have subsequently challenged Dan-

gerfield's thesis, emphasizing other factors responsible for the demise of the Liberal Party, such as the divisions created by World War I. Despite the voluminous literature germane to the collapse of Liberalism, few studies provide more than a perfunctory examination of the relationship between the Ulster crisis and the transformation of the British party system.

Patricia Jalland depicts the Ulster dilemma as a primary factor in the decline of Liberalism. After surveying the evolution of the Liberal Party's commitment to a resolution of the Irish problem, Jalland focuses on the years 1911-1914. She argues that Britain's last Liberal Government failed to exploit a rare historical opportunity. The interval between the passage of the Parliament Act of 1911 and the growth of popular support for the Irish republican movement was, according to Jalland, perhaps the only period in modern times when neither constitutional restraints nor resentments nurtured by violence inevitably doomed compromise over Ulster. The Liberals' 1912 Irish Home Rule Bill, however, failed to provide special provisions for Ulster. Prime Minister Herbert Asquith receives much criticism for his tendency to prevaricate and rely on ambiguity and delay and his failure to understand the depth of Ulster's grievances. The passage of time, asserts Jalland, rendered political compromise more difficult, and the Curragh fiasco in March 1914 eliminated the option of coercing Ulster with force. The Liberals' 1911-1914 preoccupation with Ireland, she concludes, impeded land, fiscal, and educational reforms, thus prompting the working class to vote Labour after the Great War.

Based on private papers, official publications, *Hansard*, newspapers and periodicals, and contemporary published memoirs, *The Liberals and Ireland* invites appreciation for its empirical mastery and sophisticated analysis. Its two major theses—that the Liberals mismanaged Home Rule and that this failure contributed significantly to the party's decline—appear irrefutable. Moreover, Jalland shades her generalizations with subtle distinctions. Perhaps, however, the somewhat circumscribed scope of the monograph merits a mild caveat. By giving only summary treatment to the impact of Liberal preoccupation with Ulster on the working class, Jalland's conclusion risks the intrusion of an ellipsis. Moreover, more attention to public opinion and other parties would have provided additional context for evaluating Asquith's options. Nevertheless, specialists will find *The Liberals and Ireland* an important contribution to British political history.

WILLIAM M. SIMONS
Oneonta State University College
New York

ANTHONY KOMJATHY and REBECCA STOCKWELL. *German Minorities and the Third Reich.* Pp. vii, 217. New York: Holmes & Meier, 1980. $26.50.

What but existential despair overwhelms the reflective observer as he contemplates the shifting fortunes of Central Europe's German minorities between the two world wars. Deeply committed to the preservation of their language, their schools, their cultural agencies and traditions, these ethnic enclaves found themselves held hostage by the ebb and flow of Germany's policies and the circumstances of their host nations. In each of the five countries examined (Czechoslovakia, Hungary, Poland, Romania, and Yugoslavia), political tensions and economic distress gave rise to both domestic and foreign policies that impinged dramatically on the lives of the German groupings. Local assessment of the policy consequences either strengthened or weakened the hands of German ethnic leadership in the direction of insularity or cultural ecumenism. As a by-product, a generation gap along ideological lines erupted among the German ethnics, with the younger advocating an *erneurung* (revitalization) favorable to radical nazification, while their elders tended to resist.

The stereotype of the *Auslandsdeutche* has long been in need of erasure. This minority has been maligned as a traitorous cabal, completely disaffected, and engaged in nefarious schemes of subversion in behalf of the Reich. The present study, supported by an ample Notes section, clearly establishes that these Germans were neither homogeneous nor disloyal, whether considered among their lands of settlement or within a given sovereign state. They differed considerably in terms of traditions, religion, material conditions of existence, political outlook, local institutions, and contiguity with the German frontier. Thus, within Poland, the circumstances and preoccupations of the German minority in Silesia differed from those in Danzig; within Romania, the experiences and lifestyle of the Transylvanian Saxons were not those of the Bessarabian Germans.

When economic conditions deteriorated, the vulnerable Germans not infrequently were the victims of discriminatory practices as in Sudetenland and Silesia—and, to be sure, in governmental employment everywhere. Beyond that, magyarization and polonization policies often left the German populations hard-pressed, while religious rivalries among Roman Catholics, Orthodox, Protestants, and others of the dominant elements precipitated repercussions in the German communities. Komjathy and Stockwell repeatedly demonstrate how the Reich manipulated the ethnic Germans in furtherance of its own imperialist designs, yet abandoning them to local maltreatment whenever immediate interests so dictated. Indeed, in most instances, the German minorities aspired to autonomy at most, for they lost economic advantages when incorporated within the Reich. The eventual Nazi *gleichgeschalt* meant not only new hardships, but, for many heartbroken parents, the recruitment of their sons for the Waffen SS.

The present study gives rise to a transcendent ethical concern, addressed superficially in the Introduction and Conclusion. This is the issue of cultural pluralism versus split loyalties in a multiethnic state. The prevailing implicit assumption is that the German minority (and other ethnic enclaves), if unmolested with respect to language, schools, press, and cultural associations, would voluntarily assimilate in the course of several generations. An alternative hypothesis of a durable multiethnic society, whose components freely associate for political and economic advantage in a federal union, is virtually ignored. Along with this theoretical defect is the absence of a set of maps, indispensable to all but initiated Central Europeans. Finally, the structuring and style of composition is such that the reader is introduced to an endless parade of local personalities and organizational names, compelling several rereadings to permit seeing both forest and trees. Limitations notwithstanding, the work remains a valuable contribution, in that it increases our knowledge while helping to dispel unwarranted cliches.

ELMER N. LEAR

Pennsylvania State University
Middletown
Pennsylvania

JAMES RIORDAN. *Soviet Sport: Background to the Olympics.* Pp. xiii, 172. New York: Columbia University Press, 1980. $20.00. Paperbound, $8.95.

Without a doubt one of the most successful sport systems of modern times has been that of the USSR. Since Moscow's reentry in 1952 into the modern Olympic games, its athletes have consistently been at the top of the unofficial medal and points standings. Riordan, a lecturer at the University of Bradford and a highly regarded authority on Russian sport, describes and analyzes the organizational characteristics of this prominent sport system. The timing of the publication of *Soviet Sport* is excellent, since Moscow recently hosted the first Olympics to be held in a communist nation and that event was the subject of much controversy—none of which is covered, unfortunately, by Riordan in this book.

Soviet Sport contains a detailed description of the history and evolution of the Soviet sport system since prerevolutionary days. It is an account of how young talent is discovered and nurtured, how the Soviets make very good use of a ladder incentive system to motivate athletes to strive to greater achievements, how coaches and physical educators are trained for careers in their chosen sport, how women are encouraged in the pursuit of sport, how the Soviet system balances the appeal of physical culture and elite athletic training, how the sport schools operate, and how the emphasis on research—particularly physiological and psychological—goes hand in hand with the emphases on participation and international sport success. Riordan also makes it clear that the Russians are not afraid to assert that sport has distinct functions for the sociopolitical system, such as the interpretation of success on the athletic field as concomitant with political superiority.

Interspersed throughout this rather sterile descriptive account are intimations that all is not well within the Soviet sport system. For example, there are those who feel that the emphasis on the preparation of elite athletes for world-class competition is a compromise of the socialist principles which call for the greatest good for the greatest number. On the other hand, those who support the development of the elite athlete assert that the benefits of victory in the international arena serve the worldwide cause of communism. Riordan also notes that corruption, professionalism, and violence are rampant in the Soviet system, particularly with soccer. Rarely do we hear of these discords, because all accounts of Soviet and Eastern European systems, including Riordan's other publications on international sport, depend on official statistics as the prime source of information.

Riordan's references to trouble in the system are guarded and made without commitment to pursuing their subsequent analysis. This is not surprising, since Riordan clearly believes that the Soviet sport system is without equal in terms of producing world-class athletes and a physically fit population. The centralized, nationalized system provides the best apparatus for subsidizing athletes, for integrating athletics and education, for funding research, and for providing appropriate recognition for athletes. Other countries would do well in Riordan's eyes to consider the implementation of a similar sport system.

Soviet Sport is not the best of Riordan's publications on Soviet sports. It contains neither the detail nor the scope of *Sport in Soviet Society* (Cambridge, 1977) or *Sport under Communism* (Columbia, 1979.). The fact that his discussion neglects the political ramifications of the 1980 boycott and the events leading up to the United States' withdrawal suggests that Riordan has committed a severe oversight. To overlook the boycott controversy is to deny the reader an opportunity to come to grips with a larger issue: the role of sport in international relations and diplomacy. Riordan could have enlightened us all on the true nature of this political event; his choice not to, if for any reason other than the publication time-table, represents a real disservice to the reader.

JAMES H. FREY

University of Nevada
Las Vegas
Nevada

ISSER WOLOCH. *The French Veteran from the Revolution to the Restoration.* Pp. 392. University of North Carolina Press, $27.00.

For many of us who write on French topics, there was a time when to arrive in Paris meant to climb wearily from an overcrowded Air France bus at the air terminal of Les Invalides. But I wonder how many among us had any idea that the adjacent Hotel des Invalides, after which the air terminal and metro stop had been named, had once sheltered almost 5000 retired soldiers and that it was there, and not at the Bastille, that the most decisive revolutionary action had taken place on 14 July 1789.

Isser Woloch has written a history of this institution and its inhabitants during the period that runs from the late years of the Old Regime to the Restoration, in the context of the policy of the French state toward veterans. To reconstruct this dual history was no easy task, for it required sifting through an enormous quantity of government documents, distinguishing between what was legislated or decreed and what was actually done, establishing a statistical profile of the body of revolutionary and Napoleonic veterans, then weaving the results of these various inquiries into a coherent narrative that did not give the impression of two books bound into one.

Louis XIV founded the Hotel des Invalides in 1674 as part of his effort to regiment his subjects and to create visible symbols of the grandeur of his reign. Under his successors, veterans policy in general and the management of the Hotel des Invalides in particular came to provide more examples of the inequities and inconsistencies that were discrediting the Old Regime. Pensions were given capriciously and above all to aristocratic generals as rewards for service to the crown; and a large part of the budget of the Hotel des Invalides went to the royal favorites who administered it, rather than to the retired soldiers who had been forced by their lack of means to take refuge within its walls.

Reforming ministers tried to rationalize veterans policy, in their desperate attempt to curtail state expenditure: one event went so far as to stipulate that those who chose to enter the Hotel des Invalides renounced their right to a pension for life. But it was not until the National Assembly took up the issue and rendered public the abuses that had been tolerated under the Old Regime that any real progress was made.

No brief summary can convey the complexity or the richness of this important and well-crafted book. Woloch has penetrating observations to make on a host of issues, ranging from the regional provenance of French veterans to surgical procedures in the revolutionary and Napoleonic armies. (One reason so many rank-and-file soldiers survived to claim officer status at the Hotel des Invalides was because of the frequency and skill with which French army doctors hacked off limbs.) While clearly sympathetic to the egalitarian legislation passed by the Convention in 1793, Woloch is careful to note the political conformism of most veterans who appear to have adapted as easily to the Restoration as they had to the Revolution itself. He also perceives that, regardless of the regime that administered it, the Hotel des Invalides guaranteed a longer and more comfortable life to those poor and disabled veterans who were able to gain admittance to it than they could secure for themselves in the world outside. From this mundane—but far from irrelevant— point of view, the French state's experiment in social welfare must be pronounced a success; and so much Isser Woloch's book, which is an example of the new social history at its best.

ROBERT WOHL

University of California
Los Angeles

UNITED STATES

ERIC FONER. *Politics and Ideology in the Age of the Civil War.* Pp. vi, 250. New York: Oxford University Press, 1980. $15.95.

This collection of essays by a distinguished young historian is a welcome antidote to the fragmentation and overemphasis on quantification which have been in vogue recently among social historians. Foner deplores "the divorce of social from political and intellectual history" and feels that "the retreat from the analysis of political ideas deprived social history of the larger context which alone could have imparted to it a broader meaning." Thus his major concerns in these essays are "subjects eclipsed in recent years: politics and ideology. . . . These essays, written between 1965 and 1980, reflect not simply a common interest in the causes and consequences of the American Civil War, but my ongoing desire to reintegrate the political, social, and intellectual history of that period."

The first two essays deal with the origins of the Civil War. Here Foner discounts the "new political history," with its emphasis on quantification and its own brand of cultural determinism, and the "modernization" theory as tools for explaining the Civil War. He insists that the basic causes of the Civil War are to be sought in the growing social and ideological cleavage between North and South, due to the institution of slavery and the developing political structure which gradually underwent realignment until the political parties became the instruments of sectional ideology.

A second group of essays is devoted to ambiguities in the antislavery movement. In "Abolition and the Labor Movement in Ante-bellum America" and "Racial Attitudes of the New York Free Soilers," Foner ably explores the attitudes and interests of free white labor, Hunkers, Barnburners, Free Soilers, and others. Support for the antislavery movement came slowly and from widely different motives, and in the end the price of that support was the deletion of any commitment to equal rights for blacks from the political platform.

A final group of essays, under the heading of "Land and Labor after the Civil War," presents three topics: "Reconstruction and the Crisis of Free Labor"; "Thaddeus Stevens, Confiscation, and Reconstruction"; and "Class, Ethnicity, and Radicalism in the Gilded Age: the Land League and Irish America." The first two papers are excellent portrayals of the abiding effects of race and class on the outcome of Reconstruction. The third is somewhat tangential to the main theme of the book, but it is a perceptive analysis of the rise and decline of Irish-American radicalism in the face of certain political, economic, and ideological realities in nineteenth-century America.

It is to Foner's credit that his work does not bear the stamp of any particular "school" of historical writing. He has broad interests in intellectual and political history and a good supply of common sense. As a sociologist who has always considered history an essential ally of sociology and who, 45 years ago in an essay entitled "Patterns of Race Conflict," tried to extract from historical documents certain patterns of Southern thought and behavior, I am quite ready to applaud Foner's kind of social history.

GUY B. JOHNSON

University of North Carolina
Chapel Hill

JAMES L. GARNETT. *Reorganizing State Government: The Executive Branch.* Pp. viii, 243. Boulder, CO: Westview Press, 1980. $25.00

In this volume Garnett attempts to answer three basic questions about state reorganization efforts: Why do state reorganization efforts occur? How are the organization efforts conducted? And what forms do executive branches which have been reorganized take? In order to answer these fundamental questions, Garnett systematically examines all of the reorganization efforts that have occurred between 1900 and 1975. This effort results in a data base of 151 state executive branch reorganizations.

In order to determine why reorganization attempts are made, he first reviews the various theoretical perspectives such as the "socioeconomic determinants" approach, the reorganization as "diffused innovation" perspective, and reorganiztion as "adaption to modernization." Garnett provides us with a synthesis of the various theoretical perspectives in order to "take advantage of the collective strengths and compensate for the shortcomings of individual approaches." The third chapter is an excellent overview of the various forms that reorganization efforts have taken, as well as an attempt to determine what criterion should be used in evaluating reorganization efforts. These related tasks lead him to develop a new "modified typology" which is used to determine the degree of reorganization reform. In Chapter 4 Garnett discusses the various strategy alternatives available to bring about the adoption and implementation of reorganization decisions. He presents the reader with an excellent comparative overview of the different methodologies used to study executive reorganization as well as his own systematic quantitative approach. Garnett's work meets all the standards for vigorous, value-free research and leads him to conclude that state reorganization adoption has been more successful in New England, the Southeast and the Far West. Further reorganization efforts occur more frequently at an early point in a governor's administration, and most attempts are made by statute rather than by constitutional amendment or executive order. The last chapter presents Garnett's conclusions and guidelines for practitioners to translate the research findings into policy actions and a thoughtful section devoted to possible directions for further research.

This volume is well-written, concise, and filled with pertinent findings. Garnett has done a commendable job in bringing conceptual order and empirical systematic research methodologies to the study of reorganizing the executive branch of state government. This work should result in an increased awareness of a field of inquiry that has been long neglected. I believe it would be a valuable addition to the library of anyone interested in public administration or state and local government.

JOHN S. ROBEY

East Texas State University
Commerce

G. CALVIN MACKENZIE. *The Politics of Presidential Appointments.* Pp. xxi, 298. New York: Free Press, 1980. $19.95.

The subject of presidential appointments has not received the systematic scrutiny it deserves. When the topic is brought to the public's attention, it is because of some embarrassing political blundering on the part of the nominee or the president making the appointment. The general absence of visual conflict masks the politics behind various groups competing to have their candidate receive coveted nominations. On rare occasions, the Senate may stonewall action on an executive appointment; rarer still are instances of outright rejection, as in the cases of Nixon's nominations of Judges Carswell and Haynesworth to the Supreme Court.

In this small volume, Mackenzie, a political scientist and former research analyst for the Commission on Administrative Review in the Congress, provides a theoretical framework for analyzing executive appointments to top federal positions. Mackenzie wields an even, and deft, sword in attacking the procedural and political problems in the appointment-confirmation process. The narrative is less than exhilarating, but the message of the book is well taken: we are rewarded with competency "when careful and persistent efforts are made to select appointees of the highest quality," but all too often cronyism and secrecy characterize the appointment process.

Mackenzie divides the scope of his analysis into four sections. The first two, the selection and confirmation processes, receive most of the attention. The latter two—the informal role played by other than Senate and presidential participants in the process, and the conclusion on reforms to depoliticize the process—round out his analysis. Mackenzie is concerned primarily with aggregate appointments from 1945 to 1978 in the executive branch and to the major regulatory commissions. During this period, the average tenure of cabinet secretaries was a little more than two years, making the appointment process a continual one. One of the strengths of this book is Mackenzie's focus on the evolution of the appointment process from a relatively personal process in our early years to the computerized and institutionalized system of today.

Appointments are as valuable and important for patronage and symbolic reasons as for their impact on the direction of public policy. Truman, Johnson, and Ford inherited their administrations in a nonelective manner. Eisenhower, the first Republican president in 20 years, was unfamiliar with Washington politics and relied on several key individuals for appointment advice. Kennedy's emphasis on youth and vigor brought some new faces to government, but his slim 1960 election victory required that he reward the party faithful as well. Mackenzie credits Kennedy with establishing a modern rational process for recruitment and Johnson with taking a personal interest in the selection process. It was during Kennedy's brief tenure that the impetus to recruit women and minorities into top-level

positions began. Yet, it was Johnson who made these first significant appointments. As the war in Vietnam flew out of control, Johnson appointed people whose loyalty to the president was the prime qualification. The conflict between competency and loyalty begs for elaboration here; how far can a presidential appointee go in his public or private dissent from presidential policy? Nixon, Ford, and Carter draw low marks from Mackenzie both because of the manner in which their appointments were handled and because of the lackluster qualifications of some of the appointees. The problems faced by Ford were heightened as they became linked with Watergate, a hostile Senate, and his own election ambitions. Carter's appointees fell short of the high standards publicized by the president, and his appointments did not strengthen his ties with Congress or broaden the base of his own political support.

Mackenzie's involvement in the appointment process has provided him firsthand insights and anecdotes to illustrate his assessments. He is neither pessimistic nor optimistic about reform of the procedures involved in selecting top-level executives. What he finds most disquieting is an absence of any consensus on the criteria that qualify individuals for presidential appointments. Lacking these, the Senate's formal confirmation powers are diluted. This system may discourage presidents from nominating political hacks to important policy positions, but it does not enhance the selection of the most competent.

JOHN H. CULVER

California Polytechnic State
    University
San Luis Obispo

SOCIOLOGY

CARY CHERNISS. *Professional Burnout in Human Service Organizations.* Pp. xviii, 295. New York: Praeger Publishers, 1980. $18.95.

Professional burnout is described as a syndrome of many negative factors.

These include stress, strain, boredom, self-doubt, dissatisfaction, insecurity, disappointment, and frustration. Burnout is usually experienced by some newly trained professionals who are employed in large bureaucratic public agencies, frequently during their first professional appointment. This phenomenon is the focus of the study. The consequence of professional burnout is identified as a change in attitude and characterized by many negative behaviors.

The study was conducted in four different work settings: poverty law, mental health, high school teaching, and public health nursing. The major findings include the following.

The primary sources of stress result in self-doubt, or questioning by the professional about his own competence.

Coping with stress and the social changes inevitably follow. For example, the professional school indoctrinates the student with a strong sense of ideology geared to the needs of the client and the responsibilities for rendering services. The advent of professional burnout transforms these idealistic goals to an emotional detachment from the client and, second, into a pronounced self-interest or "me-first" attitude.

Data have been collected by the effective use of life-histories. These case studies exemplify how the researcher was able to compare, contrast, analyze, and reach some sound conclusions about the problem of burnout.

Finally, the project explores many other factors: the work setting, the career orientation process, and the quality of the novices' private life, including the role and personality of the individual professional.

Granted that an inquiry has basic limitations, several critical questions must be raised about this study.

Why was this research limited to new agency workers? In view of the fact that burnout is a highly contagious occurrence, why not extend the testing to other staff members who worked for five years, or to older workers who devoted 10 to 20 years to the public agency? Furthermore, probing this problem with supervisors might have revealed some deeper causal factors.

The author recommends the hiring of paraprofessionals and part-time workers as a preventive measure to burnout. This can be challenged on two counts. First, the dropout rate among "paras" is extremely high. Second, agency personnel frequently resist the employment of semitrained professionals. Therefore, it might have been very useful to compare and contrast the burnout factor of the total agency staff members. How much alienation and frustration did paras experience in comparison to other full-time professionals? How do various types of agency staff members cope with burnout?

The study overlooks the conflict between the ideal and reality. This conflict has been expertly investigated by Gunnar Myrdal in *The American Dilemma*. Cherniss overlooks the significance of idealism. The professional school and its faculty indoctrinate the students in an ideology that structures the rationale, purpose, and existence of a profession and the need for understanding the client in all of his weaknesses. Remove the idealism and focus the curriculum only on the harsh realities of practice, and who on earth would be attracted to embark as a professional in this field?

On the positive side, Cherniss has structured a study in a new area, the existence of professional burnout. The ingenious use of the personological interview and the imaginative approach to the development of life-histories provides an excellent research method for the testing of small samples where statistical manipulations are not feasible. Therefore, while cause and effect are not applicable, this study provides a sizable array of hypotheses that can pave the way for many other studies about the tensions and anxieties that scorch so many novices working in the human services organizations.

MARTIN E. DANZIG
ASE and City University
    of New York,
    Kingsborough
    Brooklyn

JAMES S. LARSON. *Why Government Programs Fail: Improving Policy Implementation.* Pp. xiv, 124. New York: Praeger, 1980. No price.

Defining "failure" broadly as "any significant shortcoming in a government program" that brings about subsequent changes in the law or its implementation, Larson proceeds to examine five federal programs in his quest for useful generalizations about what to do when failure occurs, his "Rules of Program Failure" under various conditions.

The subtitles of his five core chapters indicate the range of his research: "Energy, Failures in Regulation"; "Health, Failures in Federal Partnership"; "Education, Failures in Equal Opportunity"; "Environment, Failures in Implementation"; and "Public Housing, Failures and Economic Forces." Each of these chapters includes a historical review and a summary table presenting by dates the main changes that occurred under the parallel column headings of "Policy," "Problem," and "Remedy." There are also good bibliographies.

In his final chapter, Larson brings together his findings about appropriate and inappropriate responses to failures encountered. These are further summarized in a table entitled "Reasons for Program Failure," where he groups them under four headings: (1) poor implementation, (2) intergovernmental complexity, (3) unrealistic or vague goals, and (4) economic changes. He cites four types of reaction to these failures, sometimes good and sometimes bad: (1) new guidelines and regulations, (2) partial reorganization, (3) new legislation and amendments, and (4) termination.

For instance, new guidelines and regulations were seen as an appropriate response when the reason for failure seemed to have been poor implementation, as in oil policy and Medicaid ("good rules can improve implementation in most cases"). Partial reorganization of administration gets approval where there have been "implementation failures caused by poor organization" or where agency activities have been poorly coordinated due to intergovernmental complexity, as with Medicare and environmental protection programs. New legislation and amendments seem to be appropriate responses when the goals have been unrealistic or vague—a common occurrence—or there have been major economic changes (as in energy and education). Termination is deemed appropriate only in two cases: where goals are so unrealistic or vague that a "new legislative mandate cannot clarify goals," or where economic changes have been substantial (the best illustration is the Arab oil embargo of 1973 and the consequent termination of mandatory import quotas).

Governmental social programs in this country have been the result of the failure of the private sector to provide desired (Larson calls them "necessary") goods and services in desired amounts for certain groups. Lacking that solution, these groups and their sympathetic supporters turn to the government for help. If majority support can be found in Congress, laws are passed which state goals and set up agencies to try to achieve them following certain general guidelines. Initial funding follows, but is often proved inadequate. Rare is the program that fully achieves its goals within its first budget (an example is military cost overruns).

When failure is perceived in the private sector because of lack of profit motivation and then in the public sector because of lack of funds, what next? Some say, "Fund it more lavishly," as in recent Democratic administrations strongly oriented toward "social welfare" and Reagan toward "defense." Republicans in 1981 would be more likely to terminate and leave voids unfilled, cut budgets and talk about eliminating "unnecessary spending, waste, and fraud," or make block grants to the states and cities and let them make up the deficits while struggling to set up their own goals, administrative structures, and necessary staffs.

Larson's slender volume should prove valuable reading for both scholars and those on the firing line: legislators, administrators, and advocacy groups. No program design or execution will ever be perfect, but it seems wise to be aware of the pitfalls social pioneers encountered and tried to meet.

ROBERT B. PETTENGILL

Tampa
Florida

CHIN CHUAN LEE. *Media Imperialism Reconsidered: The Homogenizing of Television Culture.* Pp. 276. Beverly Hills, CA: Sage Publications, 1980. $22.00. Paperbound, $9.95.

Lee's book is an attempt to apply Lenin's concept of imperialism as well as the later day left-wing concept of dependency to a study of TV in nations. There is much confusion as to subject matter and method throughout the book.

Lee begins by discussing the Marxist and neo-Marxist theories of development (imperialism and dependency). He argues that a few also believe in cultural imperialism. He then discusses capitalist theories but with one exception: he only discusses those aspects of non-Marxist theory that apply to the concept of imperialism. The exception is a discussion of Daniel Lerner's early work relating communication, urbanization, literacy, and development, which work he rejects.

Lee then quotes Stinchcombe to the effect that the reason two studies could prove opposite points of view was that the studies were at diffrent levels of analysis. Lee then defines media imperialism as "(1) television program exportation to foreign countries; (2) foreign ownership and control of media outlets; (3) transfer of the 'metropolitan' broadcast norms and institutionalization of media commercialism at the expense of 'public interest'; and (4) invasion of capitalistic world views and infringement upon the indigenous way of life in the recipient nation." I kept looking for his promised "level" of analysis discussion, and then it turned out that the four parts

to the definition were meant to be the four levels of analysis. This is not my understanding of the level of analysis problem.

There were some general statistics and discussion on the four points, followed by short chapters on Canada and Taiwan, where the four points were discussed in a little greater detail. There was a chapter which recapitulated Lee's arguments and presented several conclusions. This was followed by an epilogue which discussed Red China. At all times the author tried to take a position between what he poses as the Marxist and the capitalist position, but never is he very persuasive.

This book never has a clear objective. If it is a critical view of Marxist literature, why are not such theoretical problems as the conflict between Marx's stages of development and dependency addressed—surely all developing nations are not capitalist. The review of "capitalist" literature, except for Lerner, is much too brief and too specialized. If Lee is concerned with development, why study Canada? Is Canada an underdeveloped country? He seems at times to want to develop a theory of his own, but the four-part definition of media imperialism is as close as he comes. The work often tries to present data as if empirical tests of theories were intended, but clearly Lee is not conversant with empirical methodology. For example, at one point he talks of differences in the significance of certain percentages, but no statistical tests of significance of his data are ever presented. Important terms like "homogenization" are never defined.

An introduction by Elihu Katz says the book makes four points:

—Canada must struggle to avoid domination by the United States.
—Taiwan's TV seems to be developing a mass culture that is neither American nor Chinese—not even Taiwanese.
—Taiwan makes indigenous programs that outdraw "Kojak."
—Foreign ownership of media, except ad agencies, is not often successful.

I suppose Professor Katz is correct, but why was the theoretical superstructure imposed?

O. ZELLER ROBERTSON, Jr.
Saginaw Valley State College
University Center
Michigan

ANDREW ROLLE. *The Italian Americans: Troubled Roots.* Pp. 222. New York: Free Press, 1980. $14.95.

Italian-Americans have made substantial economic and social progress since the great migrations to the United States from Italy at the turn of the century. This progress is now coming into full public view through the efforts of social and intellectual historians, following the path paved a half-century ago by Giovanni Schiavo; the work of the large national organizations; scholarly magazines such as *Italo Americana*; and the general circulation magazine *Attenzione.*

Andrew Rolle, Cleland Professor of History at Occidental College and one of the most distinguished of the historians of the Italian-American experience, does not attempt to deal with external realities in this new volume; he feels that far too much attention has been given to economic and political success. He is concerned, rather, with the inner motivations of the Italian immigrants and the effects of these motivations on their children. He proposes a new kind of ethnic history, one based on modern psychoanalytic insight.

The result is decidedly mixed. His approach is certainly useful and deserves further attention. But Rolle takes far too negative a view of the inner history of Italian-Americans. He correctly notes that many of the immigrants were poor, but further concludes that there was deep alienation, a lack of self-identity with these immigrants. He writes that many immigrant children have tried to reject their heritage; that on the whole there has been much psychic failure and unhappiness on the part of Italian-Americans. Naturally, there have been failures with Italian-

Americans, but Rolle's book presents far too negative a picture.

It can be persuasively argued that, on the whole, contemporary Italian-Americans are as much, if not more, in touch with their cultural heritage as any other ethnic group—that they have successfully adapted to their new land while in many cases respecting their old.

Further, Rolle's concepts about the Italian immigrants themselves seem totally out of focus. These immigrants realized in most cases that substantial success would not necessarily be their lot, but they also knew their sacrifices were paving the way for their children. This certainly was not lack of self-identity, but rather the highest form of realism and self-sacrifice. For these immigrants were sacrificing their material well-being so that their children would prosper. They gambled and their children are now reaping the profits of their gambling.

The inner motivations of these immigrants do deserve study, but preferably a more balanced and thorough analysis than is presented in this volume.

JENO F. PAULUCCI
The National Italian American
Foundation

ECONOMICS

DAVID CAPLOVITZ. *Making Ends Meet: How Families Cope with Inflation and Recession.* Beverly Hills, CA: Sage Publications, 1979. $16.00. Paperbound, $7.95.

In 1973 and 1974, after a long period of continuous prosperity, the United States experienced a serious economic reversal. Conventional macroscopic measures of consumer prices, wages, and unemployment indicated that on the average consumers were less well off than they had been. Caplovitz's study is addressed to questions of the actual impact of inflation and recession on American families. What groups were most affected? How did families cope with an upsurge in rates of inflation coupled with high rates of unemployment? Results are reported of a survey conducted in 1975 of nearly 2000 families in the New York, Atlanta, San Francisco, and Detroit metropolitan areas. A stratified sampling procedure was used to assure that comparisons could be made of the poor, blue-collar workers, white-collar workers, and the retired. Data are reported in two forms: tabulations of responses to forced-choice questions and quotations of responses to open-ended questions.

To measure how families were affected by inflation, Caplovitz used two approaches. "Objective inflation crunch" was measured by responses to summary questions about trends in family economic fortunes in the preceding few years. "Subjective inflation crunch" was measured by means of an 11-item scale seeking to tap suffering that respondents attributed to inflation. Overall, disadvantaged groups, that is, the poor and the minorities, reported subtantially greater negative inflation impacts on both measures. A more complex pattern emerged among retired persons, who reported high economic impact but only modest suffering, which Caplovitz attributes to their modest expenditure requirements.

Respondents were asked how they cope with inflation. Caplovitz identifies five broad coping strategies to organize their answers: income-raising, curtailment of expenditures, self-reliance and bargain hunting, sharing with others, and use of credit. Overall, those most affected by inflation and recession reported most use of each of the strategies.

Subsequent chapters are concerned with the impact of inflation on family life and mental health. Caplovitz also examines the potential mitigating roles of receipt of public assistance and homeownership. In general those who claimed to have been affected greatly by inflation also tended to report marital and mental strain. Homeowners were less affected by inflation than renters. No evidence was found to suggest that receipt of public assistance reduced the subjective, family, or mental impacts of inflation and recession. In fact, recip-

ients of various forms of public assistance reported that they were worse off than others on all impact measures.

Conspicuously absent are both a review of previous research on consumer response to changing economic conditions and a discussion of methodological issues. The lack of attention to methodology is particularly regrettable, since the technical problems that underlie research in this arena are highly complex and the approaches which Caplovitz used are full of problems. Not only does Caplovitz rely on ex post facto data, but he bases the analysis on gross summary respondent opinions. As already indicated, "objective inflation impact," for example, simply reports general respondent impressions. The validity of such reports as information on changes in actual family fiscal conditions is open to question. No attention is given to the psychometric properties of the scale used to measure subjective inflation impact. Simply on the basis of an examination of item content, there is reason to question the degree to which suffering attributable to inflation is measured.

The statistical techniques used also deserve comment. Caplovitz relies extensively on tabulations and cross-tabulations for presentation of quantitative data. There are no tests of statistical significance. No use is made of more sophisticated approaches to multivariate analysis. The minimally technical approach to the handling of quantitative data is useful in making the survey results accessible to a general audience, but some acknowledgment might have been made of the limitations of the approach. Caplovitz does not consider that cross-tabulation may be an incomplete basis for controlling for the effect of a variable. The absence of an effort to separate effects of income from effects of race/ethnicity is another example of the superficial approach to analysis.

Caplovitz's own ideologically based opinions regarding the causes of inflation and the alleged indifference of policymakers to its consequences are unwelcome distractions.

The study is most useful in identifying patterns of individual and family response to economic adversity. Implicitly Caplovitz's research makes a case for a much more carefully designed longitudinal inquiry that develops a more concrete and detailed picture of the responses of American families to continuing economic challenges.

FRANCIS G. CARO

Institute for Social Welfare Research
New York

RICHARD A. EASTERLIN. *Birth and Fortune: The Impact of Numbers on Personal Welfare.* Pp. xii, 205. New York: Basic Books, 1980. $11.95.

Easterlin's thesis is that, largely because of federal policy, "in the post World War II economy the success of a generation's members may be crucially affected by how numerous they are." Generation is a group of persons born in a particular year—using birth *rate*, not absolute numbers.

The implication of the thesis is significant: "that the general deterioration in the American economy and society . . . will abate over the next two decades as the presence of numbers diminishes." But the improvement will not last: "The U.S. economy may be embarked on a self-generating cycle of around forty years." The explanation, the opposite of standard demographic "age structure" explanations, is that the baby boom children have small families because of difficult economics (economic expectations *relative* to one's parents), and the baby bust children have larger families because of better economics—less competition, less unemployment, faster advancement, and so on.

Easterlin suggests some points worthy of serious consideration: the adverse psychological impact of generation size "possibly fostered a state of mind more responsive" to the alienation of Vietnam and Watergate; this "Vietnam generation" will continue to be a prime target for alienation despite "superficial conformity"; and personal and social problems may result more from the generational size cycle than from larger societal changes. His specific attention to women's work is welcome (generation size has an even greater impact on women), but some specific conclusions are surprising. Easterlin analyzes explanations opposed to his; he notes that his methodology might be criticized as too simple but that he covers a large number of subjects and documents substantial historical differences; and he notes that his projections are based on only one cycle that goes back to 1940, but argues that this is better than extrapolations from only a half-cycle—that is, since 1960.

Easterlin's thesis is straightforward and consistent with theories (relative income, anomie, relative deprivation, cognitive dissonance) in other disciplines, but it is a minority view and is likely to be provocative. His suggestion that government use policies which alter the composition as well as the level of labor demand is worthy of serious consideration.

The accumulation and analysis of data is creative, comprehensive, and impressive. The clear and effective writing style is an enjoyable bonus. *Birth and Fortune* is strongly recommended for both professional and nonprofessional readers, and for readers from diverse disciplinary interests.

ABRAHAM D. LAVENDER

University of Miami
Florida

GEORGE GILDER. *Wealth and Poverty.* Pp. xii, 306. New York: Basic Books, 1981. $16.95.

This book, receiving wide attention, is a melange of characteristics and topics. The writing is colorful, effective, exaggerated, full of extraordinary turns of phrase—and obviously polemical. It is *not* a conventional, scholarly volume. The argument too often is overdrawn, picking fights across a range of issues where it would seem they need not be picked. In many instances the analysis displays imaginative perception, but it also belabors the obvious, beating some dead horses that both liberals and conservatives have grown tired of attacking—for example, command planning and the current welfare "mess." Gilder attacks macroeconomics, static equilibrium theorizing, the "myths" of discrimination, big business as the source of innovation and dynamics in contemporary economy, "Keynesianism," GNP accounting, and the welfare state in general and in particular.

Gilder, however, deals with more than the negative. He presents cogently the case for "supply-side" economics. He is an optimist looking to the future, espousing a dynamic view of resources and of human capabilities. Resources can expand under the onslaught of technology, science, and entrepreneurial drive. A nation's wealth is not in its oil, land, gold, or even machines, but in a culture that energizes risk-taking, hard work, and innovative imagination. Both government and big business have their roles, to play, but the mainspring of future innovation resides in creative, uncertainty-bearing, small-firm entrepreneurs.

Liberal oxes are gored in exaggerated fashion, but Gilder's analysis also is bound to displease conventional conservatives. Monetary explanations of inflation are regarded as incomplete and mechanistic. Government at all levels is seen as here to stay. When all is said and done, he accepts much of the mixed economy. It is a quibble about degree rather than kind. This is not to suggest that quibbles are unimportant. Incremental shifts and adjustments in institutions and policies within a free economy seeking justice are crucial matters for analysis and debate. Public-sector expansion of the last two decades, in Gilder's view, needs to be absorbed and improved—as well as pruned. Federal deficits, though not something to brag about, are clearly the lesser of evils in many circumstances. Safety nets need to be lowered and revised rather than eliminated. He calls for child allowances across all incomes rather than AFDC payments, and urges a continuation of in-kind subsidies, though at more austere levels.

The book has a distinctive flavor with the argument that, at bottom, dynamic capitalism is an altruistic exercise in giving rather than exchange. The rationale for this unexpected idea is that enterprisers with new investments risk without guarantees and are presumably cognizant of the desires of others in society. The lack of quid pro quo associated with entrepreneurial profit-seeking suggests for Gilder that investment is more than exchange. He tries to build from the concept of reciprocity in Stone Age culture to the idea that investment in new products and processes is basically altruistic. At least this discussion will fuel a fair amount of debate! Interestingly, more economists are probing the role of altruism and trust in a market economy. Gilder's book is part of that development.

His presentation ends with emphases on mystery and religious comment about faith, hope, and love. It is ironic that one who wears, in the eyes of many, the label of "new right" closes with poetry composed by that "liberal-minded" political thinker, Reinhold Niebuhr. Though this book is overblown and excessively argumentative in many places, it addresses crucial questions with perception and imagination. It is a worthy addition to ongoing debate about the nature of the American economy and where it goes in the future.

HAROLD L. JOHNSON
Emory University
Atlanta
Georgia

BEN W. HEINEMAN, Jr., and CURTIS A. HESSLER. *Memorandum for the President: A Strategic Approach to Domestic Affairs in the 1980's.* Pp. xxv, 404. New York: Random House, 1981. $17.95

Both of these young authors served as assistant secretaries in President Carter's administration. Both were well educated as Rhodes Scholars and as law clerks of the U.S. Supreme Court. They have had valuable experience in the executive branch of our national government. They have written with great skill and thorough knowledge of the responsibilities, the limitations, the need for flexibility in the efforts to solve the problems that will confront the president in the 1980s.

Indeed, Heineman and Hessler write that the "modern presidency is an institution in trouble." The people expect the president to govern the nation and compel him to struggle fiercely to control the flow of events in his administration. He must make decisions affecting each measure that comes to the Oval Office. Furthermore, to achieve success he is forced to create a new coalition in his attempt to solve the problems that were acted upon.

Confronted with all of these limitations and obstacles, the president must pursue a strategic method in order to maintain the proper communication between the four types of aides on whom he relies for advice. These advisors were divided into those who promote economic activity, those who champion domestic affairs, those who specialize in budget and management, and those whose interests were in politics only.

Franklin Roosevelt was a "skillful manipulator and a brilliant interpreter." Whereas the economic conditions, we are told, aided Roosevelt, the economy will be a thorn in the side of the presidents of the 1980s. The national budget has changed greatly in "size, composition and nature" in the last 40 years.

The crux of the presidency is the Cabinet. The president must depend upon the complete loyalty of his Cabinet at all times. He must determine which issues will be his alone, which he will share with his secretary, and which he will leave to the secretary alone.

Heineman and Hessler argue convincingly that the Chief of Staff is the most important of the president's domestic appointments. The Chief of the White House Staff must keep the relations among the president's executive advisors operating coherently and harmoniously. It is essential that the president establish close relations with the chairman of the Board of Governors of the Federal Reserve System.

Yet another area of communicative relations necessary for the president are those with Congress. More than one president has seen his legislative program defeated because of congressional opposition within his own party. Carter presented bills for liberal programs to Congress but lacked the necessary leadership to get them enacted by Congress even though his party had a majority in each house. Furthermore, near the end of Carter's administration he oscillated between the politics of governing and that of renomination and reaped the worst of both.

Finally, this memorandum warns the president that between 1981 and 1985 there will be no easy solutions to the economic problems, the tax issue, the wage-price measure, the monetary, the fiscal—all of these problems are closely intertwined. No policy adopted will affect all of them well.

This is a unique, though significant book. Its use will far outlast the 1980s. It should be read by all Americans who are interested in the survival of the American institution—the presidency.

GEORGE C. OSBORN

Gainesville
Florida

WILLIAM J. NORDLUND and R. THAYNE ROBSON. *Energy and Employment.* Pp. xiv, 128. New York: Praeger Publishers, 1980. $17.95.

DUDLEY J. BURTON. *The Governance of Energy.* Pp. xxx, 395. New York: Praeger Publishers, 1980. No price.

U.S. energy policies have evolved haphazardly and are frequently inconsistent, outdated, insufficient and inequitable. Therefore, Nordlund and Robson state that their purpose is not to solve all the problems but merely to stimulate more research and thought by calling attention to them. Discussion is concentrated on (1) energy production industries and (2) all other industries in which energy is a factor. The former includes mining, processing, refining, generating, transporting, distributing, and selling energy as electric power, as liquid fuels, or as heating materials, and their employment components.

Employment by the total energy sector is not too sizable but is certain to expand rapidly. However, the employment impacts of conservation policies and practices are not fully examined as their net effect on the use of energy is not yet clear. It is suggested that in the near future solar, shale, geothermal, wind, wave, and fusion technology as abundant and inexpensive sources of energy will actually play a minor role in the nation's energy strategy as they are only now in the experimental stage. Most of the new production employment is expected to come from the rapid expansion of electricity production and the development of new coal mines.

Energy is also a factor in the production/consumption process of almost all other industries. It is more pervasive than any other factor input. A study thereof is required as the availability and price of the energy will determine location and employment. However, the most powerful modeling and analytical techniques are inadequate to determine employment figures with any degree of exactitude. Nevertheless, as the national energy policy emerges, Nordlund and Robson hope for a more orderly, efficient, and equitable transition into a new energy era, one of probably higher energy prices, a shift to alternative sources, increased conservation, and improved efficiency.

Burton's volume is an outgrowth of a 1977 international conference, "The Energy Crisis—Implications for Governance." It is oriented toward a general audience of students as well as engineers, economists, and other specialists. It aims to examine energy problems from the perspective of social and political affairs.

Similarly, he declares that the various problems must first be recognized and dealt with responsibly, then plans must be made to manage them so that at least some measure of social welfare, social justice, and environmental quality may be achieved with respect to the use of energy resources.

Energy governance is especially essential in the United States since the control of world energy markets through the control of world oil has passed from its hands. However, Burton concludes that such energy governance will not be likely until fundamental changes are made in the structure and orientation of our bourgeois society. Social development and urban form and lifestyle must first be taken into account before economic costs and risks, even though these have important implications for political values and relationships; it cannot be the other way around.

The future may mean higher prices for energy to consumers, streamlining of regulatory processes, and reestablishment of competitive pricing in basic fuels. Yet this makes sense only if there is sufficient control over individual corporate behavior to ensure that there is not just a "private" government in these energy resources. Conservation and belt-tightening among consumers while suppliers of energy maintain growth and increase profitability is a contradiction. On the other hand, regulation of prices and supplies without consideration for social well-being and social justice is equally perverse.

Both these books appear to follow a similar socialistic path, and their conclusions are not too dissimilar; they are well written and well printed and contain fine bibliographies, appendices, notes, and especially excellent charts figures. They provide food for thought in the present energy crisis and deserve serious attention.

ALBERT E. KANE
Washington, D.C.

ALEX RADIAN. *Resource Mobilization in Poor Countries: Implementing Tax Policies.* Pp. xxiv, 266. New Brunswick, NJ: Transaction Books, 1980. $19.95.

This book examines problems faced by underdeveloped countries in generating tax revenues. The evidence for this research was obtained through Radian's personal interviews with tax agency officials of Jamaica, Trinidad, Thailand,

and the Philippines. The useful portion of this book reports on these interviews.

The middle four chapters report on specific problems those countries face in their process of taxation. The chapters deal, respectively, with the general problem of insufficient manpower and funds, with audit and collecting procedures, and with the sham of enforcement. The problems resulting from (U.S.A.I.D.-inspired) reforms in Jamaica and Trinidad, and from discretionary collection procedures in all countries, are illuminating. Unfortunately, in his attempt to demonstrate a general theory, Radian obscures the problems of any one of the four countries by frequent use of intertwining references.

The rest of the book is less satisfactory. The opening sections, which review the literature on theory of taxation, are difficult to follow. The results of various studies are not well integrated, and in one case a table is reprinted from another study with an incomprehensible explanation of the reported data. The discussion of tax level determinants assumes that governments choose taxes to maximize their own revenues. ("When a country does not use up its entire taxable capacity it is because it did not try hard enough.") This assumption is not standard in public finance theory.

Stylistic and editorial factors make this book difficult to use. Cliches and rhetorical questions abound. Graphs are mislabeled. At one point Radian discusses the need for decentralizing the audit and collection process into local communities. This is followed immediately by this claim: "Though decentralization is indispensable, it amplifies the already acute problems of personnel and expertise without substantially improving the nation's revenue situation." One wonders what the previous pages were about. Finally, many assumptions and conclusions are supported by "charts" comprised of lists of quotations from other authors who previously said the same thing. This practice is distressing to one trained in the scientific method, especially in cases where one is fairly sure statistical evidence is available.

This book offers some useful insights into the problems and operations of tax revenue departments in poor countries. The cost of gaining these insights, however, is high.

GEORGE T. McCANDLESS, Jr.
Dartmouth College
Hanover
New Hampshire

MICHAEL REICH. *Racial Inequality: A Political-Economic Analysis.* Pp. xii, 345. Princeton, NJ: Princeton University Press, 1981. $22.50. Paperbound, $6.95.

I wonder where to start the discussion of such an excellent and most necessary book. Whether one is a scholar, policymaker, or interested general reader, Michael Reich has provided a veritable banquet of information and analysis from which to feast.

The basic item placed before the reader is the question of how significant the market economy U.S. style has been in generating and perpetuating racial inequality. The conclusions arrived at by Reich's skillful and exhaustive analyses are very important, particularly as we begin to partake of Reaganomics with its private market selective relishes and desserts. Contrary to the wishes of George Guilder and Milton Freidman, to name perhaps the newer and elder purveyors of private market delights, the evidence suggests that since 1865 the market place has been a primary source of racial inequality. Wrapped within the profit-seeking motives of competitive markets, "racial inequality will be reproduced, not eliminated . . . because capitalists gain from it." And furthermore, "it appears that the redistributive effect of racial inequality is concentrated among the very rich; they are the prime beneficiaries."

A delightful aspect of this well-integrated presentation is that Reich provides readers with a complete selection of explanation for social inequality. In my opinion, Reich legitimately displays the arguments of each major theory and demonstrates their respective shortcomings; this is particularly the case with respect to the dominant neoclassical paradigm. Reich has the best of it when he suggests that "economic analysis is better served by a class conflict and power analysis than by the individualistic and market-oriented analysis of neoclassical economies."

Chapter 4 is not easily digested because of its utilization of most necessary but complex econometric analyses of who benefits from racial inequality. The findings are so important—"white workers lose from racism while rich whites, capitalists, and a few privileged workers benefit"— that the slow going seems a small price to pay for such tantalizing stuff.

In Chapter 6 Reich demonstrates his versatility by shifting from the complex mathematical and theoretical analyses of earlier courses to the narrative mode of comparing case studies. He focuses here on "class differences among whites and class conflict variables" and "the relation between governmental activity and economic forces." The conclusions arrived at through these blends are that racial and class inequality result from dividing workers through appeals to racial symbols and state intervention in support of such behavior—both calculated to erode worker solidarity and bargaining power.

If anyone seriously anticipates a lessening of inequalities as a result of Reagan-Stockman economic conjuring, then one does not wish to dine at the best table—that set by Michael Reich in this book. *Racial Inequality* is a veritable cerebral-humanitarian gourmet's treat!

CARL F. PINKELE
Ohio Wesleyan University
Delaware

W. W. ROSTOW. *Why the Poor Get Richer and the Rich Slow Down.* Pp. xvii, 376. Austin: University of Texas Press, 1980 $19.95.

Utilizing a background of economic history and theory, Rostow presses his view of the inadequacy of neo-Keynesian and neoclassical theory as a foundation for present economic policy in a series of eight essays.

A third of the book is devoted specifically to a review and unification of long-cycle theories, including a mathematical model of the Kondratieff cycle. In addition, Rostow integrates long-cycle theories with actual cyclical upswings and downswings in the last two centuries in several other chapters. This is consistent with his work in *The Process of Economic Growth* (1953) and emphasized more recently in *The World Economy: History and Prospect* (1978). Kondratieff is regarded as the father of the idea that capitalist economies are dominated by cycles—mainly in prices— that reach lengths of 50 years. Rostow reviews the limited success Schumpeter and Kuznets had in the development of the "appropriate theory of long waves." According to Rostow, the lesson to be learned from previous efforts is that long cycles can be explained only on the basis of a general, disaggregated dynamic theory of production and prices. This theory would involve an explanation and interrelationship of technological change, relative price movements, and population trends.

The late 1970s and early 1980s are in reality the fifth Kondratieff upswing. Rostow suggests that the theoretical constructs of conventional economists have handicapped economic expansion by continuing to orient policy toward the manipulation of aggregate demand and thus have failed to reorganize the main body of economic theory to deal with a supply-oriented world. Rostow remains optimistic that there are ample opportunities for productive investment in the future in resource-related fields, which could prevent periods of stagnation of real income and chronic unemployment.

What is needed, according to Rostow, is an elusive "dynamic disaggregated mode of analysis" linked to neo-Keynesian income analysis and to the imperatives of balance-of-payments equilibrium. By sacrificing elegance for realism, the complex process behind the flow of technology can be considered substantially endogenous. He stresses the need to bring science, invention, and innovation within the bounds of economic analysis, as forms of investment.

In an examination of energy and economic development, Rostow maintains that present energy-economy models fail to capture the major features of the energy problem faced by the United States and other OECD nations. Energy-economy model builders should retreat from long-run, full-employment equilibrium models and concentrate on the period immediately ahead when OPEC production will reach its limits of capacity. In the long run, he sees no foreseeable physical limit on U.S. energy supplies such as coal, shale, and advanced forms of nuclear power. Unfortunately, Rostow has relied entirely on projections of the CIA and DOE, which have often proven inaccurate in the past. Additional short-run conventional sources of energy and conservation were not fully considered.

Rostow finds that the stagnation of the OECD economies has resulted from a failure of investment to recover to levels characteristic of the great boom of the 1950s and 1960s, when a powerful expansion of investment was based on an increase in real income borne by low commodity prices. Large investments in the future are needed in energy production and conservation, water development, transportation, and raw material development.

In a chapter entitled "Money and Prices," Rostow launches an attack on the monetarist view of the economy. He criticizes the use of the quantity theory of money as the matrix for business cycle analysis, the consistency of the monetary cycle with business cycle behavior, and concludes that there is no empirical foundation for a monetary

shock theory as an initiating force in business expansions. His criticism of the cyclical application of the quantity theory rests on the failure to bring into play essential elements of long-term price trends and business cycles: changes in technology, industrial capacity, infrastructure, and the supply of foodstuffs and raw materials.

In the last two chapters he calls for cooperation to solve many economic problems. A North-South (developed and developing nations) conference should be convened on a functional basis to deal with common objectives faced by both sets of nations. However, as long as the OECD nations concentrate on policies aimed at full employment through the manipulation of neo-Keynesian monetary and fiscal policy, the chances for new directions are limited. While Rostow admits that the proposal for such cooperation may be unrealistic, circumstances dictate that the attempt be made. Rostow also calls for cooperation domestically by relating the text of remarks made to Argentine students in 1965 in which he calls for a social compact in society to maintain full employment without inflation. Such a compact would involve the setting of wages in accordance with productivity increases and setting prices in reference to the real cost, not in anticipation of inflation. (This is essentially the Kennedy-Johnson administrations' wage and price guideposts of the 1960s.)

The book offers many interesting avenues to observe contemporary economic problems from a historical and modern perspective. While Rostow may not have come up with viable alternative theoretical constructs in every case, he has demonstrated the need to reassess current economic theory and policy.

RUSSELL BELLICO
Westfield State College
Massachusetts

OTHER BOOKS

AFANASYEV, V. G. *Marxist Philosophy: A Popular Outline*. Pp. 400. Moscow: Progress Publishers, 1980. $8.00.

ALPERT, GEOFFREY P., ed. *Legal Rights of Prisoners*. Pp. 280. Beverly Hills, CA: Sage Publications, 1980. $22.50. Paperpound, $9.95.

ANDERSON, BARRY F. *The Complete Thinker*. Pp. x, 278. Englewood Cliffs, NJ: Prentice-Hall, 1980. $12.95. Paperbound, $4.95.

ARMSTRONG, J. D. *Revolutionary Diplomacy: Chinese Foreign Policy and the United Front Doctrine*. Pp. 251. Berkeley: University of California Press, 1981. Paperbound, $6.95.

ARONOFF, MYRON J. *Ideology and Interest: The Dialectics of Politics*. Pp. vi, 217. New Brunswick, NJ: Transaction Books, 1980. $29.95.

ASANTE, MOLEFI KETE and AB-DULAI S. VANDI, eds. *Contemporary Black Thought: Alternative Analyses in Social and Behavioral Science*. Pp. 302. Beverly Hills, CA: Sage Publications, 1980. $20.00. Paperbound, $9.95.

ASBELL, BERNARD. *The Senate Nobody Knows*. Pp. ix, 466. Baltimore, MD: Johns Hopkins University Press, 1981. $6.95.

BOWMAN, LARRY W. and IAN CLARK, eds. *The Indian Ocean in Global Politics*. Pp. xi, 260. Boulder, CO: Westview Press, 1981. $25.00.

BRAX, RALPH S. *The First Student Movement: Student Activism in the United States During the 1930s*. Pp. 121. Port Washington, NY: Kennikat Press, 1981. $17.50.

BREZHNEV, LEONID I. *Peace Detente Cooperation*. Pp. xiii, 197. New York: Plenum Publishing, 1981. $15.00.

BRIDGEMAN, JON M. *The Revolt of the Hereros*. Pp. vii, 184. Berkeley: University of California Press, 1981. No price.

BRODY, DAVID. *Workers in Industrial America: Essays on the 20th Century Struggle*. Pp. ix, 257. New York: Oxford University Press, 1981. Paperbound, $3.95.

BROWN, A. LEE, Jr. *Rules and Conflict: An Introduction to Political Life and its Study*. Pp. xiv, 368. Englewood Cliffs, NJ: Prentice-Hall, 1981. No price.

BURKHART, JAMES L., SAMUEL KRISLOV, and RAYMOND L. LEE. *The Clash of Issues: Readings and Problems in American Government*. 7th ed. Pp. ix, 341. Englewood Cliffs, NJ: Prentice-Hall, 1981. $9.95.

BURKS, ARDATH W. *Japan: Profile of a Postindustrial Power*. Pp. xii, 260. Boulder, CO: Westview Press, 1981. $22.00. Paperbound, $9.50.

BURTON, WILLIAM C. *Legal Thesaurus*. Pp. xii, 1058. New York: Free Press, 1980. $35.00.

BYRNE, DONN and KATHRYN KELLEY. *An Introduction to Personality*. 3rd ed. Pp. xiv, 591. Englewood Cliffs, NJ: Prentice-Hall, 1981. $18.95.

CHAMELIN, NEIL C. and KENNETH R. EVANS. *Criminal Law for Police Officers*. 3rd ed. Englewood Cliffs, NJ: Prentice-Hall, 1980. $15.95.

CHANNAN, KRISHAN K. *The Lure of Politics*. Pp. viii, 68. Smithtown, NY: Exposition Press, 1981. $6.00.

CHERNISS, CARY. *Staff Burnout: Job Stress in the Human Services*. Pp. 199. Beverly Hills, CA: Sage Publications, 1980. $20.00. Paperbound, $9.95.

CLARK, W.A.V. and ERIC G. MOORE, eds. *Residential Mobility and Public Policy*. Pp. 320. Beverly Hills, CA: Sage Publications, 1980. $20.00. Paperbound, $9.95.

COHN, ALVIN W. and BENJAMIN WARD, eds. *Improving Management in Criminal Justice*. Pp. 160. Beverly Hills, CA: Sage Publications, 1980. $15.00. Paperbound, $7.50.

COLLARD, DAVID, RICHARD LECOMBER, and MARTIN SLATER, eds. *Income Distribution: The Limits to Redistribution*. Pp. xi, 267. New York: John Wiley & Sons, 1981. $34.95.

CROSSKEY, WILLIAM WINSLOW and WILLIAM JEFFREY, Jr. *Politics and the Constitution in the History of the United States, Vol III: The Political Background of the Federal Convention.* Pp. xii, 592. Chicago: University of Chicago Press, 1981. $27.00.

DARLEY, JOHN M. et al. *Psychology.* Pp. xvi, 652. Englewood Cliffs, NJ: Prentice-Hall, 1981. $18.95.

DAVISON, W. PHILLIPS and LEON GORDENKER, eds. *Resolving Nationality Conflicts: The Role of Public Opinion Research.* Pp. xiv, 242. New York: Praeger Publishers, 1980. $18.95.

DECHANT, EMERALD. *Diagnosis and Remediation of Reading Disability.* Pp. xiv, 512. Englewood Cliffs, NJ: Prentice-Hall, 1981. $17.95.

DONALDSON, ROBERT H., ed. *The Soviet Union in the Third World: Successes and Failures.* Pp. xiv, 458. Boulder, CO: Westview Press, 1981. $25.00. Paperbound, $12.00.

DUNKERLEY, DAVID and GRAEME SALAMAN, eds. *The International Yearbook of Organization Studies.* Pp. vi, 249. Boston: Routledge & Kegan Paul, 1981. $37.50.

DUNN, WILLIAM N. *Public Policy Analysis.* Pp. xii, 388. Englewood Cliffs, NJ: Prentice-Hall, 1980. $18.95.

Education and Community Self-Reliance: Innovative Formal and Non-Formal Approaches. Pp. 223. Geneva, Switzerland: UNICEF, 1981.

EHRLICH, STANISLAW and GRAHAM WOOTON, eds. *Three Faces of Pluralism: Political, Ethnic and Religious.* Pp. ix, 219. England: Gower Publishing, 1980. $18.95.

ELMAN, NATALIE MADORSKY with JANET H. GINZBERG. *The Resource Room Primer: Teaching Techniques for Developing, Implementing, or Improving Resource Room Programs.* Pp. xviii, 296. Englewood Cliffs, NJ: Prentice-Hall, 1981. No price.

EMBER, CAROL R. and MELVIN EMBER. *Cultural Anthropology.* Pp. xviii, 397. Englewood Cliffs, NJ: Prentice-Hall, 1981. $13.95.

FERMAN, LOUIS A., ROGER MANELA, and DAVID ROGERS. *Agency and Company: Partners in Human Resource Development.* Beverly Hills, CA: Sage Publications, 1980. $6.50.

FIELD, GEOFFREY. *Evangelist of Race: The Germanic Vision of Houston Stewart Chamberlain.* Pp. 565. New York: Columbia University Press, 1981. $25.00.

FINGER, SEYMOUR MAXWELL. *Your Man at the U.N.* Pp. xx, 320. New York: Columbia University Press, 1980. $26.50.

FLORY, THOMAS. *Judge and Jury in Imperial Brazil 1808-1871: Social Control and Political Stability in the New State.* Pp. xv, 268. Austin: University of Texas Press, 1981. $19.95.

FLOTO, INGA. *Colonel House in Paris: A Study of American Policy at the Paris Peace Conference 1919.* Pp. ix, 374. Princeton, NJ: Princeton University Press, 1981. $16.50.

FREEDMAN, JONATHAN L., DAVID O. SEARS, and J. MERRILL CARSMITH. *Social Psychology.* 4th ed. Pp. xiv, 686. Englewood Cliffs, NJ: Prentice-Hall, 1981. No price.

FRIEDMAN, PHILIP. *Roads to Extinction: Essays on the Holocaust.* Pp. xiii, 610. New York and Philadelphia: Jewish Publication Society of America, 1980. $27.50.

FULTON, RICHARD M. *The Revolution That Wasn't: A Contemporary Assessment of 1776.* Pp. viii, 247. Port Washington, NY: Kennikat Press, 1981. $19.50.

GODSON, ROY, ed. *Intelligence Requirements for the 1980's: Counter Intelligence.* Pp. ix, 339. New York: National Strategy Information Center, 1980. $7.95.

GORTNER, HAROLD F. *Administration in the Public Sector.* Pp. xiii, 413. New York: John Wiley & Sons, 1981. $16.95.

GEORGES-ABEYIE, DANIEL E. and KEITH D. HARRIES, eds. *Crime: A Spatial Perspective.* Pp. xii, 301. New York: Columbia University Press, 1980. $25.00.

GRENVILLE, J.A.S. *A World History of the Twentieth Century, Vol. 1. 1900-45: Western Dominance.* Pp. 605. Totowa, NJ: Barnes & Noble Books, 1981. $32.50.

GRUMM, JOHN G. and STEPHEN L. WASBY, eds. *The Analysis of Policy Impact.* Pp. xii, 209. Lexington, MA: D. C. Heath, 1981. $23.95.

HARDACH, KARL. *The Political Economy of Germany in the Twentieth Century.* Pp. xiii, 247. Berkeley: University of California Press, 1981. $5.95.

HAWLEY, AMOS H. *Urban Society: An Ecological Approach.* Pp. xi, 383. New York: John Wiley & Sons, 1971. $14.95.

HEISS, JEROLD. *The Social Psychology of Interaction.* Pp. ix, 358. Englewood Cliffs, NJ: Prentice-Hall, 1981. $14.95.

HINCKLEY, BARBARA. *Outline of American Government.* Pp. xi, 304. Englewood Cliffs, NJ: Prentice-Hall, 1981. $8.95.

HUANG, QUENTIN K.Y. *Pilgrim from a Red Land.* Pp. xxxvii, 119. Smithtown, NY: Exposition Press, 1981. $8.00.

HUDSON, ROBERT B., ed. *The Aging in Politics: Process and Policy.* Pp. xi, 294. Springfield, IL: Charles C. Thomas, 1981. $27.50. Paperbound, $19.75.

JOHNSON, DAVID W. *Reaching Out: Interpersonal Effectiveness and Self-Actualization.* Pp. xi, 308. Englewood Cliffs, NJ: Prentice-Hall, 1981. $13.95. Paperbound, $9.95.

JOHNSON, EDWIN S. *Research Methods in Criminology and Criminal Justice.* Pp. viii, 418. Englewood Cliffs, NJ: Prentice-Hall, 1981. $17.95.

JONES, THOMAS E. *Options for the Future: A Comparative Analysis of Policy-Oriented Forecasts.* Pp. xxii, 348. New York: Praeger Publishers, 1980. No price.

JUREIDINI, PAUL A. and R. D. McLAURIN. *Beyond Camp David: Emerging Alignments and Leaders in the Middle East.* Pp. xxii, 197. Syracuse, NY: Syracuse University Press, 1981. $18.00. Paperbound, $8.95.

KALE, NDIVA KOFELE. *Tribesmen and Patriots: Political Culture in a Poly-Ethnic African State.* Pp. xvi, 359. Washington, DC: University Press of America, 1981. $22.00. Paperbound, $12.75.

KAMENKA, EUGENE and ALICE ERHSOON TAY, eds. *Law and Social Control.* Pp. ix, 198. New York: St. Martin's Press, 1981. $22.50.

KANN, ROBERT A. *A History of the Hapsburg Empire 1526-1918.* Pp. xiv, 646. Berkeley: University of California Press, 1981. $10.95.

KLEIN, MALCOLM W. and KATHERINE S. TEILMAN, eds. *Handbook of Ciminal Justice Evaluation.* Beverly Hills, CA: Sage Publications, 1980. $39.95.

KONTOS, JOAN FULTZ. *Red Cross, Black Eagle.* Pp. 216. New York: Columbia University Press, 1981. $17.00.

KRAMER, FRED A. *Perspectives on Public Bureaucracy.* Pp. x, 240. Englewood Cliffs, NJ: Prentice-Hall, 1981. $7.95.

KWEIT, MARY G. and ROBERT W. KWEIT. *Concepts and Methods for Political Analysis.* Pp. vi, 374. Englewood Cliffs, NJ: Prentice-Hall, 1981. Paperbound, $12.95.

LaFEBER, WALTER. *America, Russia and The Cold War 1945-1980.* Pp. xiii, 334. New York: John Wiley & Sons, 1980. $7.95.

LANDELS, J. G. *Engineering in the Ancient World.* Pp. 224. Berkeley: University of California Press, 1981. $15.75. Paperbound, $4.95.

LEVITAN, SAR A. *Programs in Aid of the Poor for the 1980s.* Pp. viii, 159. Baltimore, MD: Johns Hopkins University Press, 1980. $11.00. Paperbound, $3.95.

LEWIS, EUGENE and FRANK ANECHIARICO. *Urban America: Politics and Policy.* Pp. xiii, 295. New York: Holt, Rinehart & Winston, 1981. No price.

LIPSON, LESLIE. *The Great Issues of Politics.* Pp. xix, 407. Englewood Cliffs, NJ: Prentice-Hall, 1981. $16.95.

LODGE, JULIET, ed. *Terrorism: A Challenge to the State.* Pp. xi, 247. New York: St. Martin's Press, 1981. $25.00.

LORCH, ROBERT S. *Democratic Process & Administrative Law.* Pp. 278. Detroit, MI: Wayne State University Press, 1980. No price.

LOWI, THEODORE J. *Incomplete Conquest: Governing America.* Pp. xviii, 486. New York: Holt, Rinehart & Winston, 1980. No price.

McINTOSH, ANDREW, ed. *Employment Policy in the United Kingdom and the United States.* Pp. 188. Cambridge, MA: Abt Books, 1981. $22.50.

MAIL, PATRICIA D. and DAVID R. McDONALD. *Tulapai To Tokay.* Pp. xv, 420. New Haven, CT: HRAF Press, 1980. $25.00.

MAKI, LILLIAN. *Mother, God, and Mental Health.* Pp. xii, 177. Portland, OR: Metropolitan Press, 1980. $4.95.

MATTICK, PAUL. *Economic Crisis and Crisis Theory.* Pp. viii, 227. White Plains, NY: M. E. Sharpe, 1981. $18.50.

MELTSNER, ARNOLD J., ed. *Politics and the Oval Office.* Pp. xi, 332. San Francisco: Institute for Contemporary Studies, 1981. $7.95.

MILBURN, THOMAS W. and KENNETH H. WATMAN. *On the Nature of Threat: A Social Psychological Analysis.* Pp. 148. New York: Praeger Publishers, 1980. $19.95.

MULLER, PETER O. *Contemporary Suburban America.* Pp. xii, 218. Englewood Cliffs, NJ: Prentice-Hall, 1981. No price.

MURRAY, JOHN P. *Televeision & Youth.* Pp. 278. Stanford: Boys Town Center for the Study of Youth Development, 1980. $10.00.

NANDA, VED P., JAMES R. SCARRITT, and GEORGE W. SHEPARD, Jr., eds. *Global Human Rights: Public Policies, Comparative Measures, and NGO Strategies.* Pp. xi, 318. Boulder, CO: Westview Press, 1981. $30.00.

NAPOLI, DONALD S. *Architects of Adjustment: The History of the Psychological Profession in the United States.* Pp. 176. Port Washington, NY: Kennikat Press, 1981. $20.00.

NELSON, DANIEL, ed. *Local Politics in Communist Countries.* Pp. 230. Lexington: University Press of Kentucky, 1981. $17.50.

NEWBERG, PAULA R., ed. *The Politics of Human Rights.* Pp. ix, 287. New York and London: New York University Press, 1981. $22.50. Paperbound, $9.00.

NEWTON, KENNETH. *Balancing the Books: Financial Problems of Local Government in West Europe.* Pp. vii, 218. Beverly Hills, CA: Sage Publications, 1980. $22.50. Paperbound, $9.95.

NORTHRUP, JAMES P. *Old Age, Handicapped, and Vietnam-Era Antidiscrimination Legislation.* Pp. xiii, 263. Philadelphia: University of Pennsylvania, The Wharton School Industrial Research Unit, 1980. No price.

PALMER, DAVID SCOTT. *Peru: The Authoritarian Tradition.* Pp. xvi, 134. New York: Praeger Publishers, 1980. $19.95.

PLUNZ, RICHARD, gen. ed. *Housing Form and Public Policy in the United States.* Pp. 248. New York: Praeger Publishers, 1980. $29.95.

POOLE, ROBERT W., Jr. *Cutting Back City Hall.* Pp. 224. New York: Universe Books, 1981. $12.50. Paperbound, $5.95.

PRICE, BARBARA RAFFEL and PHYLLIS JO BAUNACH, eds. *Criminal Justice Research: New Models and Findings.* Pp. 144. Beverly Hills, CA: Sage Publications, 1980. $15.00. Paperbound, $7.50.

PRIOVOLOS, THEOPHILOS. *Coffee and the Ivory Coast: An Econometric Study.* Pp. xv, 218. Lexington, MA: D. C. Heath, 1981. No price.

RAE, GWENNETH and THOMAS C. POTTER. *Informal Reading Diagnosis: A Practical Guide for the Classroom Teacher.* Pp. vi, 218. Englewood Cliffs, NJ: Prentice-Hall, 1980. $14.95. Paperbound, $10.95.

RAELIN, JOESPH A. *Building a Career: The Effect of Initial Job Experiences and Related Work Attitudes on Later Employment.* Pp. xiii, 178. Kalamazoo, MI: W. E. Upjohn Institute for Employment Research, 1980. $7.00. Paperbound, $4.50.

REAGAN, MICHAEL D. and JOHN G. SANZONE. *The New Federalism.* New York: Oxford University Press, 1981. $3.95.

ROBLEE, CHARLES L. and ALLEN J. McKEOHNIE. *The Investigation of Fires.* Pp. xiii, 201. Englewood Cliffs, NJ: Prentice-Hall, 1981. $13.95.

RODGERS, HARRELL R., Jr., and MICHAEL HARRINGTON. *Unfinished Democracy: The American Political System.* Pp. 625. Glenview, IL: Scott, Foresman, 1981. $14.95. Paperbound, $12.95.

ROSE, RICHARD, ed. *Electoral Participation: A Comparative Analysis.* Pp. 354. Beverly Hills, CA: Sage Publications, 1980. $25.00.

RUSTOW, ALEXANDER. *Freedom and Domination: A Historical Critique of Civilization.* Pp. xxix, 716. Princeton, NJ: Princeton University Press, 1981. $35.00.

RUSSELL, JAMES. *Marx-Engels Dictionary.* Pp. xxv, 786. Westport, CT: Greenwood Press, 1980. $19.95.

SAMUELS, WARREN J. and A. ALLAN SCHMID, eds. *Law and Economics: An Institutional Perspective.* Pp. vii, 268. Lawrence, MA: Kluwer Boston, 1980. $18.50.

SCHAFFER, KAY F. *Sex Roles and Human Behavior.* Pp. vii, 392. Englewood Cliffs, NJ: Prentice-Hall, 1981. $14.95.

SCHWARTZ, THEODORE, ed. *Socialization as Cultural Communication: Development of a Theme in the Work of Margaret Mead.* Pp. xviii, 250. Berkeley: University of California Press, 1980. Paperbound, $6.95.

SELZNICK, PHILIP. *TVA and the Grass Roots, A Study of Politics and Organization.* Pp. xiii, 274. Berkeley: University of California Press, 1980. $17.50.

SILVERMAN, PHYLLIS R. *Mutual Help Groups: Organization and Development.* Pp. 144. Beverly Hills, CA: Sage Publications, 1980. $8.00.

SIMPSON, GEORGE EATON. *Religious Cults of the Caribbean: Trinidad, Jamaica and Haiti.* Pp. 347. Puerto Rico: Institute of Caribbean Studies, 1980. $12.00.

SOMOGYI, STEFANO. *Introduzione alla Demografia.* Pp. 429. Palermo, Italy: University of Palermo, 1979. No price.

STROMSDORFER, ERNST W. and GEORGE FARKAS, eds. *Evaluation Studies Review Annual, Vol. 5.* Pp. 800. Beverly Hills, CA: Sage Publications, 1980. $35.00.

SULLIVAN, JOHN L., ed. *Discriminant Analysis.* Pp. 71. Beverly Hills, CA: Sage Publications, 1980. Paperbound, $3.50.

SWEET, JOHN JOSEPH TIMOTHY. *Iron Arm: The Mechanization of Mussolini's Army, 1920-1940.* Pp. xxi, 217. Westport, CT: Greenwood Press, 1980. $25.00.

THROWER, RAYNER. *The Pirate Picture.* Pp. x, 171. Totowa, NJ: Rowman and Littlefield, 1980. $12.50.

TOCH, HANS, ed. *Therapeutic Communities in Correction.* Pp. xiv, 216. New York: Praeger Publishers, 1980. $23.95.

TORRANCE, THOMAS F., ed. *Belief in Science and in Christian Life: The Relevance of Michael Polanyi's Thought for Christian Faith and Life.* Pp. xvii, 150. New York: Columbia University Press, 1980. $12.00.

TRESTER, HAROLD B. *Supervision of the Offender.* Pp. xii, 336. Englewood Cliffs, NJ: Prentice-Hall, 1980. $15.95.

TUDJMAN, FRANJO. *Nationalism in Contemporary Europe.* Pp. 293. New York: Columbia University Press, 1981. $21.50.

TULLOCK, JOHN H. *The Old Testament Story.* Pp. xiii, 433. Englewood Cliffs, NJ: Prentice-Hall, 1980. $14.95.

VAGO, STEVEN. *Law and Society.* Pp. xi, 372. Englewood Cliffs, NJ: Prentice-Hall, 1981. No price.

VALENTA, JIRI. *Soviet Intervention in Czechoslovakia, 1968: Anatomy of a Decision.* Pp. xii, 208. Baltimore, MD; Johns Hopkins University Press, 1981. $12.95. Paperbound, $5.95.

VANDENBOS, GARY R. *Psychotherapy: Practice, Research, Policy.* Beverly Hills, CA: Sage Publications, 1980. $20.00. Paperbound, $9.95.

VERNON, RAYMOND and YAIR AHARONI. *State-Owned Enterprise in the Western Economies.* Pp. 203. New York: St. Martin's Press, 1981. $27.50.

VIGUERIE, RICHARD A. *The New Right: We're Ready to Lead.* Pp. 191. Falls Church, VA: Viguerie Company, 1981. $8.95.

WALLIMANN, ISIDOR. *Estrangement: Marx's Conception of Human Nature and the Division of Labor.* Pp. xxiv, 195. Westport, CT: Greenwood Press, 1980. $18.00.

WALTER, JAMES. *The Leader: A Political Biography of Gough Whitlam.* Pp. xix, 295. St Lucia, Queensland: University of Queensland Press, 1980. $18.00.

WANTZ, MOLLY S. and JOHN E. GAY. *The Aging Process: A Health Perspective.* Pp. xiv, 322. Cambridge, MA: Winthrop Publishers, 1981. No price.

WEISS, IRVING and ANNE D. WEISS. *Thesaurus of Book Digests: 1950-1980.* Pp. ix, 544. New York: Crown Publishers, 1981. $14.95.

WEIZMAN, EZER. *The Battle for Peace.* Pp. 408. New York: Bantam Books, 1981. $15.95.

WHISENAND, PAUL M. *The Effective Police Manager.* Pp. x, 380. Englewood Cliffs, NJ: Prentice-Hall, 1980. $16.95.

WHITE, E. G. *Final War.* Pp. 218. Phoenix, AZ: Inspiration Books, 1979. Paperbound, $1.50.

WHYTE, J. H. *Church & State in Modern Ireland 1923-1979.* Pp. xiv, 491. Totowa, NJ: Barnes & Noble Books, 1980. $32.50.

WERNHAM, R. B. *The Making of Elizabethan Foreign Policy, 1558-1603.* Pp. vii, 109. Berkeley: University of California Press, 1981. $15.50. Paperbound, $3.95.

WESSON, ROBERT G. *Modern Governments: Three Worlds of Politics.* Pp. xviii, 414. Englewood Cliffs, NJ: Prentice-Hall, 1981. $17.95.

WEST, BRUCE J. *Mathematical Models as a Tool for the Social Sciences.* Pp. 120. New York: Gordon and Breach Science Publishers, 1980. $26.50.

WOLL, PETER. *Constitutional Law: Cases and Comments.* Pp. xxxii, 923. Englewood Cliffs, NJ: Prentice-Hall, 1981. $24.50.

WOLTER, OLF, ed. *Rudolf Bahro: Critical Responses.* Pp. 237. White Plains, NY: M. E. Sharpe, 1980. $22.50.

WONG-FRASER, AGATHA S. Y. *Symmetry and Selectivity in U.S. Defense Policy: A Grand Design or a Major Mistake?* Pp. vii, 164. Lanham, MD: University Press of America, 1980. $17.50. Paperbound, $8.95.

ZALL, P. M., ed. *Ben Franklin Laughing: Anecdotes from Original By and About Benjamin Franklin.* Pp. 204. Berkeley: University of California Press, 1981. $17.95.

ZEITLIN, IRVING M. *Ideology and the Development of Sociological Theory.* Pp. vi, 330. Englewood Cliffs, NJ: Prentice-Hall, 1981. $17.95.

ZEITLIN, IRVING M. *The Social Condition of Humanity: An Introduction to Sociology.* Pp. xii, 413. New York: Oxford University Press, 1981. $11.95.

INDEX

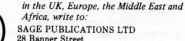